Fundamentals for a Phenomenological Study of Chemistry

by

Frits H. Julius

Published by:
 The Association of Waldorf Schools of North America
 3911 Bannister Road
 Fair Oaks, CA 95628

Title: *Fundamentals for a Phenomenological Study of Chemistry*

Author: Frits H. Julius

Original title in German: *Stoffeswelt und Menschenbildung II
 Grundlagen einer phäenomenologischen Chemie*

First Published by: Verlag Freies Geistesleben 1965

Translated from the German by D.G. Ruarus

Editor: John Petering

Copy-editor and layout: David Mitchell

Proofreader: Nancy Jane

© 2000 By AWSNA

ISBN # 1-888365-22-6

Curriculum Series

The Publications Committee of AWSNA is pleased to bring forward this publication as part of its Curriculum Series. The thoughts and ideas represented herein are solely those of the authors and do not necessarily represent any implied criteria set by AWSNA. It is our intention to stimulate as much writing and thinking as possible about our curriculum, including diverse views. Please contact us with feedback on this publication as well as requests for future work.

 David S. Mitchell
 For the Publications Committee
 AWSNA

Contents

Preface 11

I   Overcoming One-Sidedness in Contemporary
           Chemistry Teaching 13

     A New Path to Nature

     Substances and Life

     Fitness of All to All

     The Method of the True Alchemists;
         the Tria Principia and its Meaning for Us

     Three Alchemic Principles

     Man and Nature Between Light and Gravity

     Polarity of Above-Below

     The Four Elements

II   Experimentation as an Art 24

     Guidelines for Experimentation

     Apparatus

         A Demonstration Cabinet with Ventilation

         Bottom Lighting

     Examples of Experiments

         The Light Box

III   Salts, Acids, and Bases 31

Crystallization of Salts

Phenomena Occurring when Salts Dissolve

Splitting of Salt by Fire; Acid and Alkali

Salt-Formation by Adding an Acid and a Base

Transition to Chemical Equations

Weight Relationships

Oxygen, Oxidation, Burning, and Rusting

The Importance of the Approach Chosen

Reduction

Solution of Metals in Strong Acids

Electrolysis

Rounding Off the Main Lesson Block

Acid or Base Substitutions —
  Double Substitutions (Displacements)

IV   Guidelines for Teaching Chemical Formulas   53

On the Essence of a Chemical Compound and the
  Principle of Impenetrability

The Foundations of Chemical Formulas

V   Weight Ratios in Chemistry 65

The Balance and Materialism

The Balance and the Spiritual Ordering of Nature

The Spiritual Background of Materialism

Equivalent Number as a Pure Expression of
  Weight Ratios

Gas Volumes and Weight Ratios

Atomism as a Starting Point for Explaining
Natural Phenomena

VI The Process Of Dissolving—
Phenomena, Concepts, Laws 87

    The Most Important Phenomena
    The Concept of Solubility
    The Influence of Heat on Solubility
    Phenomena of Osmosis
    Explanation of Osmosis
    The Concept of Concentration
    The Concept of Dissociation

VII The Great Matter Cycles 99
    Teaching Chemistry in the 11th Grade
    Usefulness of the Element Concept
    The Periodic System of Elements in Chemistry Classes
    The Human Organism as a Key to a New System
    The Twelve Substances as a Representation of the
        Realms of Nature
    Indications for Teaching
    Oxygen 107
        Oxygen, Matter Awakened by the Sun
        Oxygen as Atmospheric Gas
        Oxygen as a Messenger from the Sun
        Oxygen and the Crust of the Earth
        Oxygen and the Hydrosphere
        Oxygen and the Tria Principia of the Alchemists
        Further Effects of Oxygen in Chemistry
        Fire as an Image of the Sun

Hydrogen 114
    Hydrogen in the Atmosphere
    Hydrogen as a Gas
    Hydrogen and Heat
    Hydrogen–the Decomposition of Living Substances
    Hydrogen and Water

Carbon 118
    Diamond and Graphite
    How Carbon Withdraws Itself from Life
    Phenomena with Combustion of Carbon
    On Carbon Dioxide
    The Importance of Carbon Dioxide and Carbon for the Life-Process in the Plant
    Carbon Between Sun and Earth
    More on Diamond and Graphite

Carbon, Hydrogen, and Oxygen 133
    Kinship and Opposition between Carbon and Hydrogen
    Compounds of Carbon and Hydrogen
    Sugar as a Balanced Substance—Harmonizing through Oxygen
    Something More About the Task of Carbon, Hydrogen, and Oxygen

Sodium and Sulfur 137
    The Natural Cycle of Sodium
    Properties of Sodium Compounds in Relation to its Place in Nature and in our Body
    The Natural History of Sulfur
    Sulfur in Protein
    A Comparison of Sodium and Sulfur Processes
    Examples for Discussing Sodium
    An Example for a Discussion of Sulfur

Potassium 151
    A Comparison with Sodium
    A Comparison with Carbon

Silicon and Silica 154
    The Occurrence of Silica in the Earth's Crust
    The Properties of Quartz in Relation to the Different Realms of Nature
    Silica in Living Organisms
    Silicic Acid and Water
    Silicic Acid and Warmth
    Silica as Oxide
    Silicon and Fluorine
    Silica Technology
    Silica and the Sun

Lime (Chalk, Limestone) 167
    Formation and Erosion of Limestone Mountains
    The Cycle of Limestone in Water
    Animal Existence as a Battle with Lime
    The Diversity of Form in Calcareous Coats, Shells, and Skeletons
    Lime in the Human Being

The Chemistry of Calcium 178
    Lime as an Alkaline Substance
    Calcium as Element
    Calcined Lime and Slaking of Lime
    Calcium Hydroxide and Carbon Dioxide
    A Few Illustrative Experiments
    Something About the Uses of Calcareous Substances
    Lime and Light—Contrast to Silicon
    Lime as Opposed to Hydrogen

Phosphorus and Magnesium 184
- On the Availability and Circulation of Phosphorus in Nature
- Phosphorus as Element—Fire and Light Phenomena
- Images of the Night Sky During Experiments with Phosphorus
- Properties of Magnesium as Element—Phenomena and Experiments
- The Occurrence of Magnesium Compounds in Earth and in Living Organisms—Contrast to Lime
- Phosphorus as Creator of an Equilibrium between Heaven and Earth and Herald of the Conscious Spirit
- The Magnesium Flame with Regard to Photosynthesis
- Phosphorus as Representative of the Stars on Earth; Results of a Comparison with Magnesium
- Magnesium Chemistry
- Phosphorus in Mythology and Industry

Nitrogen 200
- Nitrogen as a Component of the Atmosphere—The Atmosphere as Environment for Living Beings
- Nitrogen in the Soil
- Nitrogen in Protein and the Atmosphere
- Nitrogen in Nature Between Oxygen and Hydrogen
- Nitrogen Chemistry

Aluminum 212
- Aluminum as Element
- Chemical Properties of Aluminum
- Aluminum in the Crust of the Earth: Clay

Alchemical and Mythological Views of
   Pottery and Porcelain Manufacture

Nitrogen and Aluminum 217

   Polarity and Harmony in Various Realms

   Something from the History of the Production of
      Nitrogen Compounds, and the Extraction
      of Aluminum by Comparison

VIII The Structure of the Element Circle as an Image
   of the Order of Nature 226

   The Circle of Elements and the Stage of Life

   Balanced Opposites in the Circle of Elements

   Acid and Base in the Circle of Elements

   Surprising Number Relations; Air and Protein

   Metamorphosis in the Circle of Elements

   Relation to the Annual Movement of the Sun;
      Seasonal Changes and the Human Organism

IX The Halogens 232

   Fluorine, Chlorine, Bromine, Iodine—a Characterization

   Chlorine

   Fluorine

   Iodine

   Bromine

   Structural Relationships in the Halogen Group

X The Most Important Heavy Metals 250

    Introduction—a Comparison with the Substances of the Element-Circle

    Characterization of the Metallic State—Crystal and Metal

    Rusting as Expression of the Nature of Metals

Metals and Planets 256

    Tradition and Modern Research

    The Usage of Metals in Three Spheres

    Technical Usage

    Medical Purposes

    A New Realm—Metal Chemistry

Discussion of Lead and Silver 259

    Lead and Silver in the Metallic State

    Lead and Silver Chemistry

    Lead Chemistry

    Lead and Silver as Representatives of Certain World-Principles

    The Importance of Lead and Silver For Civilization

Gold 273

    Gold in the Metallic State

    Comparison of Silver to Gold

    Gold Chemistry

    The Principle Represented by Gold

The First Use of Iron and Copper as Tool-Metal 277

Iron as a Metal—The True Relation Between the Human Organism and Technology 278

Copper and Iron 281

    Iron in the Human Body and in Nature

    Iron

    The Principles Represented by Copper and Iron

    Mercury and Tin as Metals 292
        Tin
    The Chemistry of Mercury 297
        Mercury and Tin
    Quartz and Gold—The Summit in the Realm of Matter 299

XI Completing the Whole—Curriculum For Grade 12 303
    Chemistry and the Human Organism
    Chemistry as an intermediary between an Organic and Inorganic Science
    Matter and Life
    Matter and Consciousness
    Chemistry on Four Levels
    Victory over Materialism—The Task of Teaching Chemistry in Grade 12

Preface

In 1952 the pedagogical research institute of the Federation of Waldorf Schools in Germany asked me to write a book for those chemistry teachers who seek in their work to realise the pedagogical guidelines given by Rudolf Steiner. Their task is to introduce students to chemistry in such a way that they educate in the true meaning of the word. (In the introduction to the first volume this is dealt with more fully). Not only was it necessary to develop a method to teach the usual contents and concepts, but also one had to work on the basic principles for a new science. In its method, present-day science has disregarded the essence of the human being as much as possible. Now a restructuring (of the subject matter) had to found so that the nature of the human being would be at the center of this science.

To the extent that this task has been successful, the present book may be of importance to doctors, farmers, and, in general, for those people whose work lies in those areas where the interaction between substances and life forces is of central importance. This book may therefore be seen as a general introduction to the fundamentals of chemistry. I would be very delighted if many lay people, especially artists, would approach this subject, which they often find so daunting, in a new spirit. They would then experience that precisely here there are extraordinary treasures of wisdom and beauty to be found. Because our well-being depends to quite some extent on chemistry, everybody today should have some knowledge of chemistry and be able to judge on chemical issues. A science that does not incorporate the human being inevitably leads to a technology that will threaten the human being. Because of the carefully maintained arrangement of this book according to subjects, everybody will be able to find just what he needs. The fundamental discovery, on which this book is largely based, was briefly outlined in a book published in Holland nearly 30 years ago (***Goetheanistische Chemie***, part of which was published in the journal of the Waldorf School in Kassel in the thirties). This book has essentially the same structure as that first book. Its contents, however, have been considerably expanded and developed.

When I had nearly finished this work, I heard of the death of Ehrenfried Pfeiffer. Above all others he has shown me with his courses, lectures and private talks, new ways to approach nature and through that given me the key to the writing of this book.

Last I would very much like to thank Mr. Wolfgang Rudolph who has untiringly helped me with advice and in various ways. Without his help it would have been impossible for me to finish such a demanding task. I also want to thank the "Pädagogischen Forschungsstelle" of the Waldorf Association, for their generous financial aid of my work.

– Frits H. Julius
The Hague,
September 1964

I

Overcoming One-Sidedness in Contemporary Chemistry Teaching

A New View of Nature

The present form of chemistry is not the only one imaginable. On the contrary, it has presented itself in other forms which also led to important discoveries and insights, and doubtless will in the future appear in entirely new forms. One can even say that the way that chemistry is pursued nowadays—as imposing as its results might be—is nevertheless one-sided and thereby limited.

When we realize this, we are able to seek a new path which leads to other ways of regarding nature andwhy this leads to other conceptions of the world of matter. Such a path requires, also, a different structuring of chemistry teaching.

Overcoming One-Sidedness in Contemporary Chemistry Teaching

If we wish to place a correct picture of the substances and their mutual relationships around the chemistry lessons, we confront various obstacles. It occurs all too often that the writer of a textbook allows himself to be influenced by the methods which are employed in the laboratory or in industry or technology. In this area, so much has been achieved and people have exerted such great efforts that these achievements—which indeed are only acquired out of individual labor—have been involuntarily granted too great a place, and have beed regarded as having objective value.

A second difficulty is that in treating the phenomena, we often take such an excessive interest in the explanation which has been found for them, that the actual phenomena--their real, perceptible unfolding-to-completion are brought only too little to our recognition, or completely ignored. For example, if we compare the old textbooks with those of today, then we see how increasingly the theoretical considerations come to the fore at the cost of the descriptive.

A New Path to Nature

What must happen, so that a picture of nature, thus strengthened, will come into its own? To reach this goal, first of all we must always keep the relativity of our insights in view. Before the grandeur and richness of nature, they surely dwindle away. Henceforth, we ought to cherish the feeling that we are but a pupil, and above all, merely stand at the beginning of a path, which leads to the being of nature. However, as soon as we presume to have achieved any sort of final aim, she will withdraw her being.

If we put the above attitude into practice, above all we will always throw the phenomena into bold relief, present them as well arranged and as purely as possible. A phenomenon always encompasses more than can be interpreted in just a moment. We always have to make room for the possibility that later still other parts may become penetrable. Actually, along with a kind of contemplation or 'meditation' on the phenomena, one should always have the feeling: the phenomena will express something in my thoughts which she conceals from my senses. It is not I myself who explains the phenomena; rather I lend an ear to a being which declares itself in the phenomena, and will come to expression in my thoughts.

Furthermore, one is never permitted to lose sight of the fact that each phenomenon is a part of a mighty whole: the cosmos. It occurs in a definite realm and must accommodate itself to the structure and context of that sphere. We will later devote our attention in more detail to this arrangement within the world.

As much as the whole of nature is arranged around the human being as its center point, so chemistry also should be brought into close connection to the human being. Since chemistry concerns itself only with a certain domain in nature, therefore, we need to bring it essentially into connection with only a certain sheath of man's being. It is not hard to find out which sheath this is. We are able to take up and assimilate into our body only such substances as will unite with our life processes. Each substance which is assimilated offers to the life process a very definite and characteristic contact point.

Substances and Life

In this area the greatest misunderstandings arise. Above all, in the chemistry of carbon (usually called "organic chemistry") people attempt to make life superfluous, (for example, in the preparation of synthetic products and foodstuffs). They don't go on to ask how they can penetrate with understanding into the realm of the life forces. Rather, they try, with methods other than those which life ordains, to produce the same things as life, but more rapidly, or to produce things not produced by life, which are deemed more useful.

Man conducts himself in the face of life as though it were self-evidently only an interaction of inorganic materials and forces.

Despite the fact that this path produces many results, it is a path that brings us increasingly into conflict with healthy life—we have chosen a path which leads away from the truth. For, in truth, it is not possible to explain the processes in living organisms out of an interaction of matter. But, we can very well trace back the characteristics of the substances to their specific connections with life. Life is primary; our view of nature must be life-centered.

Fitness of All to All

This is strikingly expressed in the work of Henderson, *The Fitness of the Environment*.[1] He has indicated with copious material that important elemental substances such as oxygen, carbon, and hydrogen can only be understood in their chief characteristics and important relationships with a view towards the role which they achieve in a living being.

Without particular effort, one can also point this out for a whole series of different substances, most easily for the frequently occurring substances, i.e., those which are the most important decomposition products of living beings. As a consequence of this fact, the earth's body is built up in large measure out of substances whose attributes stand in a conspicuous relationship to life. This means nothing else than that life has imprinted a very clear mark on the whole spectrum of the materials of nature.

Also, it is not justified to think of life as something which once upon a time cropped up in the midst of this totality of matter, and now continuously selects the most useful materials. On the contrary, the substances are not thinkable without life, and their essential characteristics must therefore be traceable back to life. In the beginning, a unity of life and substance must have prevailed.

We have known for years that living beings play a great role in the great matter-cycles of the environment. In an impressive way, Vernadsky[2] worked this out in his book *Geochemistry*. He furnishes much material which can contribute to a great, unified view of the world of matter. We can only wish that he valued the whole as a starting point, in order to derive the characteristics of substances therefrom, rather than selecting a path which is more or less the reverse.

Henderson, on the other hand, develops the following remarkable thought sequence in his book:

"Water, of its very nature, as it occurs automatically in the process of cosmic evolution, is fitted, with a fitness no less marvelous and varied than that fitness of the organism which has been won by the process of adaptation in the course of organic evolution."[3]

Strangely, Henderson here overlooks a discrepancy in his thinking, because he focuses his gaze so closely onto the fact, neglecting that while living beings are adapted to each other and to the non-living surroundings, the dead environment is no less adapted to life. This discrepancy arises through the way he uses the word "fitness." It is easy to imagine that a living being fits itself into its environment through "adaptation" (Organic fitness). But, it is difficult to imagine (in a conventional sense) that water has "adapted" itself to life (fitness of substances to the environment).

In any case, Henderson points out an especially significant fact of nature: all its parts are attuned to one another whether we have to do with a dead substance or a living being. How can we explain this if the principle of adaptation has only a limited utility (i.e., for organic life only, albeit, treated as a mechanistic system)? Probably we will find the beginnings of a solution if we start with the development of a living being, for example one of the higher animals. In the mature animal, the organs are attuned to one another. Here, we would never find people speaking of a gradual adaptation, since we know that all that which is now differentiated into organs, was, in the ovum, an undifferentiated unity. The organs of an organism have developed out of one another, and thereby have remained attuned to one another since they belong to that whole. Could it not be so that with the whole of nature something similar occurred? Doesn't the marvelous wisdom—revealed in this atunement of all to all— tell of a primordially undifferentiated unity, from which all beings and things have developed by and by? What one person calls adaptation, would (in many cases) thus be a consequence of a development out of a common unity.

Above all, this hypothesis agrees with the results of spiritual-scientific research communicated to us by Rudolf Steiner. Just in the realm of chemistry, it is of special significance that he describes a bygone time when there was no separation between inorganic nature and living beings. Living things were then less bounded; or, alternatively, the life processes extended actively into the whole of nature. The world of substances itself was much less membered and discrete. It wouldn't have been possible to speak of chemical elements in the same sense as today. Then, nature found itself in a condition of an intermediate stage between a more spiritual and a more material form of existence. From this condition not only were living beings differentiated out, but also the chemical elements, the elemental substances, developed out of an undifferentiated wholeness, just as the organs of an animal develop from the germ cell. The phenomenon which Henderson calls "fitness of the environment" would be explained in this way.

In the introduction to the first volume of this work, we have already described that Rudolf Steiner, on the basis of his spiritual-scientific research, in every case places the human being into the center of nature. The human is not only the center point, but is also a kind of mirror image of the entire universe. Along with the differentiation between living beings and dead nature, the human being has simultaneously taken up an extract of the whole of nature into organization. As a consequence of this development, present day nature makes an impression of a grand unity which has fallen apart and has been split up; meanwhile, the human organization continually inspires the impression of a proportioned, all-embracing whole. It is possible to speak of nature as a human being fallen apart, and of the human being as a unification of the whole of nature.

In the course of our work, we will show that these basic points can become very fruitful in fashioning comprehensive principles of chemistry teaching. And just in the topics covered in grade eleven in Waldorf schools, the truth of this approach becomes clear.

The Method of the True Alchemists

Now we must ask the question: do the viewpoints mentioned above appear in our time as something totally new, or were there other people or other streams to which they can be connected? That there are tie-in points to conventional science is self-evident. However, can we also find connections when we follow other paths? This is actually so.

One of the most noteworthy phenomena in the history of science is the activity of the alchemists. Once, at a time when people were already working in the realm of physics and astronomy to develop the crystal-clear, surveyable concepts which were so characteristic for modern science, there existed a stream which at first glance makes a confusing, bewildering impression. A mixture of nature lore and mysticism is presented in these writings, and even to earnest study is not easily unlocked.

The impression of this stream takes a totally different form if we apply a particular key, which Rudolf Steiner has supplied for us. Then we can discover the most valuable traditions. In the first place, we must learn to distinguish a small group of serious researchers from a much larger number of people who more or less must be deemed dilettantes or even frauds. The serious group belongs mainly to the stream of the Rosicrucians. For them, chemistry was a not a means to achieve gold and riches, rather it was the foundation of a spiritual schooling. When they attempted to make gold, this meant primarily a striving after high moral qualities. They wanted their own being to acquire the qualities of gold.

Their work was tied into the ancient tradition of mystery schools. They referred especially to Aristotle. This is important, since at the beginning the new period of science (early Renaissance), a definite trend arose: to sever the threads from the past. Galileo vigorously attacked Aristotle, for example.

For the true alchemists, the laboratory was an altar. They did their experiments as an offering towards the gods. For our modern consciousness, this seems quite strange. However, isn't this thoroughly self evident? If we are dealing with substances and forces which have their genesis in divine creation, and if we are truly human (i.e., a divinely created being), how else can we approach such things, than with an attitude of service towards the divine? Before their work they prepared themselves through prayer. The sequence of the processes was studied, and was for them a meditation. This means that they allowed themselves to penetrate so deeply into the processes, with such devotion, that they not only acquired particular insights, but also experienced powerful formative activity, proceeding from the processes into their soul.[4]

With each important process which they contemplated, above all they asked themselves two questions: "Where and how does this take place in nature?" and, "Where and how does it take place in the human being?" This stands in accordance with an ancient tradition, which speaks of the world as macrocosm and the human being as microcosm. They considered the human being as a miniature summary of the world and saw in the world's variety and diversity an outstretched and extended human being.[5]

We will see that above all, by doing experiments in our lessons, we can gather important methodologic guidelines from this insight. Moreover, today we must always put forward a third question: "How is a process used in technology or industry?"

Three Alchemic Principles

The alchemists spoke repeatedly about a "Tria Principia: Sal, Quicksilver, Sulfur"—salt, mercury, sulfur. These not only signified certain special substances, but pointed to certain principles, which were embodied most characteristically by these substances.

Salt signifies crystallization processes, as well as everything that involves condensation and hardening of substance, and especially points to the transition from the liquid to solid state.

Mercury symbolizes everything that is mobile interaction, in transformative activity, and especially the manifold matter cycles with their play of condensation and evaporation, as represented by especially by water.

Sulfur exemplifies the process of burning, and also processes in which warmth arises and matter disappears.

If a person earnestly contemplates such thoughts, they will experience that thoughts like these can contribute a great deal towards forming a grand, imaginative picture of the whole of nature.

The alchemists didn't stop at fleeting pictures, but deepened themselves in their whole being in contemplating the world.

If they wanted to deepen themselves in the salt-principle, they did a crystallization of a salt-solution. This always awakens the impression of purity and cool beauty. In their inner devotion to this impression, they learned to look up to the sublime and rigorous divine thinking, as it reveals itself to them in the wonderful synthesis of beauty and precise form of the crystal. That which had come to expression in nature, likewise became an image in their own thinking. They strived to school their human thinking into an image of the divine as they meditated on salt crystallization.

In order to immerse themselves in the mercury-principle, they would allow a beautiful crystal to dissolve. They saw something which was initially hard, solid, and enclosed within itself, gradually become mobile and surrender itself to some other substance. Right before their eyes they saw something like the surrender of one being to another, but with a completeness which is only possible in nature. In such divine images, they schooled their feelings.

They deepened themselves in sulfur processes by igniting a fire. At first, something which was solid and bounded lay before them; then, it was destroyed, and as it rayed out in light and warmth, it passed over into an outstreaming into the surroundings. They meditated on this as a mighty image of the will. Before them stood the pure willing in preparedness, the personality totally surrendered thereby, so that a person placed themselves in the service of the Divine without reservation.

The best alchemists also practiced healing. With their "Tria Principia," they moreover had a point of entry into the physiology of the human organism. They knew that in our sense-nerve organization, substances in the head undergo a solidification (akin to a crystallization); and knew that our will-metabolic rests on a process of destruction of substance (fire), e.g., in the muscle-limb system. And, they knew that the feeling system is carried by rhythmic, harmonizing processes, concentrated in the chest.

The Tria Principia can also be exceptionally important for us, since in many respects, it provides direction. Initially, we can constantly school ourselves in the Tria Principia. Certainly, Dr. Steiner usually gave totally different methods, and I think that for most people it is not advisable to put the initial or central emphasis on alchemic methods. Nevertheless, we can achieve very definite results with it.

Furthermore, in the Three Principles we have a key to the architecture of the world, which although it takes its style from earlier times, is not out of date. In a deeply-grounded world-view, we cannot do without the Three Principles, since they are an expression of the real activity of an ordering principle. In particular, it will be difficult to find the connection between the human being and nature if we don't know this Principle, since we are an actual ordered expression of the Three Principles.

For the teacher of natural science, it is doubly important that the Three Principles be well known. They are not only a help in forming a well-rounded picture of the world, but they also give rise to an awareness of how contemplation of nature can work formatively on the human soul.

Man and Nature between Levity and Gravity

In the Three Principles we have already presented one main aspect of the organization of nature. There is yet another aspect, which manifests as a polar, but intimate relationship.

Polarity of Above-Below

If a person wishes to cultivate a view of nature, where all that we see becomes a gesture or an expression, then they must bring along an empathy for the significance of above and below. In bygone times, people had this awareness naturally; they spoke of "heaven and earth" and considered these as the background of all existence.

Through the development of the Copernican system, the relationship to "heaven and earth" became relative. A person no longer felt justified in taking themselves, their own locale, as a starting point; thus, they lost the ground under their feet and the heavens above their head. But, this is just the thing we have to regain: the capacity to start with an image of the world, just as it actually comes to meet our senses.

The human being acquires their first experience of below-above as they learn standing and walking. Whoever has taken up the contemporary way of perceiving, above all the trained natural scientist, has learned to renounce this early experience. Correspondingly, they also had to learn to run afresh, or in any event did learn to run and move themselves so that they were able to experience things further away in the realm of space. But, in order to work their way genuinely into this realm, they should practice eurythmy—an art of movement initiated by Rudolf Steiner which is derived in large measure from the finer spatial relationships between human being and the surrounding world. Here we find a domain where we can orient ourselves with certainty only if we don't just think about how it actually is, but rather also live into it with our feelings, and even develop our will as a source of insight and experience.

If people want to have a relation to the world that is free of abstraction, then they must quietly cultivate a geocentric or even "anthropocentric" point of view. Otherwise, as people immersed themselves into the processes, they would neglect the way the processes actually manifested. They would develop concepts which have too little connection with perception or are construed out of specific concepts.

Insofar as people begin with the perceptions themselves, then they will develop imaginative concepts which are closely tied to reality. In addition, they will notice very quickly how much is achieved through learning to know the spatial relationships and processes with the aid of their own bodily feelings and experience.

In eurythmy, we learn to move in such a way, that our body continuously experiences different relations to the surroundings and even to the universe. Incidentally, we develop the ability to follow everything that takes place outside the human body with our own bodily feeling. Thus, our judgment is more strongly founded in reality than in abstract concepts. Using the example of our own bodily experience, we learn how to evaluate [judge] the relationship of substances used in specific processes and their relationship to the surroundings.

When we speak of high and low in nature, it is considered from our own standpoint. We can take a further step into reality if we realize how the realm above us is mostly permeated by light and how the realm below is governed by gravity. Above and around us we find the realm where the substances are light and transparent, and where the image of the heavenly bodies arises. The nature of everything out there is such that we feel ourselves drawn to seek the periphery of the heavens. Below us, everything is dark and dense; it carries and supports our body, but also resists it.

The light in the world above draws our consciousness with it into distant widths and heights; the heaviness draws our body down into the world beneath.

When we say the words light (levity) and heaviness (gravity) we point to one of the most significant polarities in nature. The human being's poise and bodily form are determined to a great degree by this polarity.[6] With the sprouting plants we observe the roots immediately going downwards and the stem striving upwards. Animals mostly find a one-sided connection with one realm or the other.

There exists a very close relationship between the Tria Principia and the polarity of levity-gravity. Thus, we will see again and again that substance and process are related to these spheres in a characteristic way.

The Four Elements

By using the old teaching lore of the four elements, we can penetrate to a central aspect of the order of nature. Long ago, people had a very living picture of the interactions of natural forces and substances based on this teaching. This picture slowly faded away, and nothing remained but a concept of the three states of matter. The "earth element" became the dead concept of the solid state, the "water element" became the liquid state, and the air element became the gaseous state, while the warmth element degenerated to become a completely insignificant condition of matter. Dr. Steiner has indicated different ways of reviving the teachings of the four elements and even how to give them a wider meaning in modern ways, of course, that meet our present standards of precision.

In the first place, he has made possible a new understanding of the warmth element, by pointing out the progressive metamorphosis of element to element and to other strictly regular relationships between the elements.[7]

All solid substances follow gravity and are confined to the earth. This also applies to a great degree to the liquids, although the tendency is already there to leave the earth (vaporize). Gravity is largely overcome in the sphere of the air element. The play of forces in the element of warmth is, in accordance to its essence, completely opposite to gravity. It removes the substance from the influence of the earth.

The interplay of warmth forces are just as much related to light, which itself is oriented towards the periphery of the universe, as to the earth element, which is related to gravity and thus to the center of the earth.

The concept of warmth as an element has faded away as people increasingly viewed the world externally from the point of view of gravity.[8]

If we want to try and follow the development of flowering plants in order to gain a deeper insight into the activity of the life-processes, an intimate knowledge of the four elements is necessary—because the plant incorporates them into its being. And if we succeed in recognizing/knowing the activity of these four elements in the growing plant, it will help enliven the somewhat schematic picture we may have of them.[9]

In the seed the plant is earth, filled with waiting life; in the unfolding bud we first find the activity of the water element; as soon as the leaves have fully developed, carbon dioxide assimilation begins and with it an intensive interaction with air and light; with the processes in the blossom, above all the formation of pollen and seeds and the ripening of fruit, the drying, withering warmth element is active.

If we follow the development of the plant through several generations, we can have the insight that the elements can be ordered in a different way. While it develops, the plant passes through the above ascending sequence of the elements. With the falling seed, the plant life turns back again to earth, in something like a circular pattern. There is not only a linear sequence but also a circular one as well:

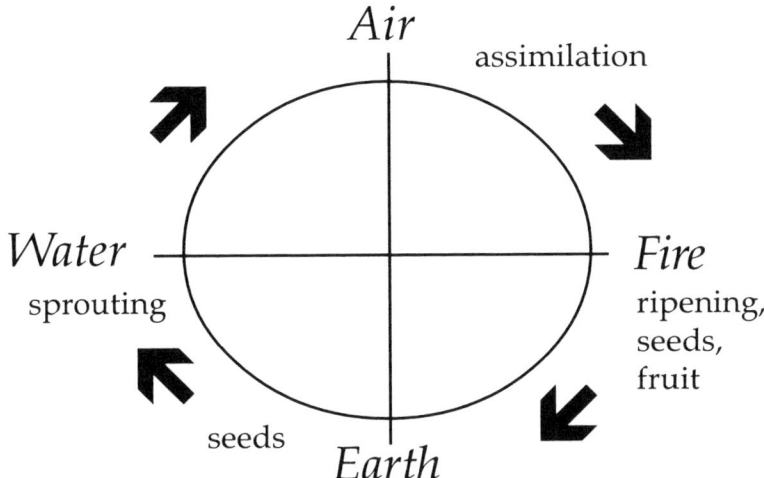

This pattern/arrangement can often be discovered repeatedly in the relationships of the beings of nature or in the sequence of natural processes. We will find this particularly important when we want to find a basis for the interconnections of the substances which we will talk about for the 11th grade chemistry block.

Endnotes
1 Lawrence J. Henderson, *The Fitness of the Environment*, New York, 1913; reprinted Peter Smith Pub. Gloucester, Mass. 1970, p131.
2 Wladmir J. Vernadsky, *Geochemistry in Selected Chapters*, Leipzig, 1930.
3 Ibid., p131.
4 Rudolf Steiner, *Rosicrucian Christianity*, Neuchatel, Sept. 28, 1911, GA 130, lecture 3.
5 Steiner, *Mystery Knowledge and Mystery Centers*, Nov.-Dec. 1923, GA 232.
6 See, *Lime and the Human Being*; p171.
7 Steiner, *Warmth Course*, GA 321, Anthroposophic Press, 1988.
8 See p40, *The Battle for the Image in Chemistry*.
9 Steiner, *Man as Symphony of Creative Word*, Lect. 7, GA 230.

II

Experimentation as an Art

With a renewal of the instructional methodology in the subject of chemistry, it is natural that the style and method of experimentation may not remain unconsidered or unaltered.

Conventional science rests in large measure on a foundation of experimental research, and one can only develop the greatest admiration for the methods which are utilized. However, if one organizes the teaching too much according to these usual methods – which are appropriate in research – we won't do the students justice. In the school there belongs, firstly, a fully developed demonstration experiment. This is something valid on its own, which must be implemented according to its own viewpoint.

In a scientific experiment, we are usually trying to detect the presence of some substance, or to prove or disprove a theory, or explain some phenomenon. So, typically we are working in realms which are new fields of inquiry, or which we want to develop further.

However, in school we are dealing above all with areas which people already know and comprehend. That which is already general knowledge to a degree must be presented in the best possible way and as impressively as possible. But above all, we ought to base all demonstrations on allowing nature herself to speak as impressively as possible.

With many usual demonstration experiments, too much emphasis is given to the theories we have developed, i.e., to human thought or technology. For example, we are often dealing with phenomena which are noticeable through strong color development; nevertheless, we tend to pay less attention to the color as such and to what it might express, but rather take it only as an indicator: this color appears, so this material is present or that process has occurred.

If we wish to show the same experiment as a demonstration-experiment, not only should we pay attention to the color itself, but also—through every sort of technique—we should allow the color to develop to a brilliant expression. There is an "art of experimentation" where in we can allow ourselves to be led entirely by nature to some degree, and place our faculties at her disposal, thereby allowing her to show herself as unhindered and directly as possible. To this end, we must learn to see how something lies in the phenomena which could be called a delicate intention or design. If we take such impressions seriously and seek to realize these "intentions," then the greatest surprises follow. We should always start from the conviction that in nature something is concealed, which can be revealed and, in a certain sense, wants to be manifest. In the preparation of an experiment, we shouldn't rest until we have the same feeling towards it as towards a plant: there, out of a jumble of green leaves, the splendor of the blossom develops. With such a style of experimentation, something like a "will towards the redemption of nature" must play a part. We will then notice that the processes become so matured and intensified [see Goethe's method] that an entire world-picture expands. In many cases, the images reflect a particular part of the macrocosm; in other cases, it appears as though a fragment of the prehistory of the earth is revived. Perhaps it is even possible, also, to make visible images of future stages of the world. However, with this we have reached a realm which goes beyond the domain of what we do in school.

In summary, perhaps we can say the following: nowadays, in doing an experiment, we usually start out from some intellectual, scientific or technical point of view. The alchemical experiments of earlier times had a strong religious background. What must be developed is the experiment as a kind of "art in practice." However, with this we naturally have to do with art in the sense of Goethe, which is at the same time a unity with science and religion.

Guidelines for Experimentation

Now, we can turn to a discussion of concrete methodology. Some guiding principles can be given, which if we follow them, will elevate the quality of the experiments to a splendid level.

The first of these principles concerns the size of the vessel—air or water container—in which we allow it to occur. To every experiment belongs a definite format, through which it is best appreciated. Usually we let the experiment occur in a flask which is too small. Nevertheless, almost always we obtain surprising activity when we increase the scale. We should have available beakers and erlenmeyer flasks of the 2-3 liter size, and in special situations even larger ones! Glass cylinders about 50 cm tall are expecially useful (often called "hydrometer cylinders"); naturally, we also use flasks of the usual size. In certain cases, smaller beakers and cylinders (12 ml) may be recommended. A variety of reagent glassware is very desirable.

Furthermore, we should set up an apparatus in which a fire can be ignited, which is clearly observable, and from which dangerous vapors can be pulled away. A glass case ventilation when placed on the demonstration table is especially effective. (We will return to this in more detail below.) Doubtless, a good exhaust duct mounted over the flames or fuming liquid will have the needed effect.

The second principle has to do with the temporal sequence of the experiment. One part of this involves each individual's tempo of manipulation and another part deals with the rounding-off, ripening, which occurs in many experiments if one lets them stand quietly for a while.

Generally, we should not work too quickly, and also should not intentionally break off the experiment too soon. For example, if we get a precipitate to form while pouring one liquid into another, it is usually best to begin with a small quantity and introduce it bit by bit. Thereby, we give the precipitate an opportunity to develop impressively. Usually, a characteristic play of colors or forms will develop in this way. If we allow the whole thing to stand overnight, usually something surprising will occur, as if nature grasped the proffered opportunity in order to develop a glorious play of color and form.

The third principle is connected to illumination. Even with the most mighty phenomena, something of the activity is lost through weak or poorly directed lighting, or an error in darkening. Therefore, we should have a strong light source available, and a way of making real darkness (thick room shades). For lighting a space, filament lamps of at least 150 watts are best; with fluorescent tubes false colors too easily arise.

The clearest experimental image is obtained if we darken the room and only illuminate the experiment itself, or just place it by a light source. Thereby we have the advantage of concentrating the attention of the students, and they also receive strong impressions at even moderate distances. In this way, we can demonstrate in front of even 150 people.

It is very advantageous, if we have special apparatus (light box) to illuminate the experiment. In doing this, experiments must be differentiated; those that require a dark background and are side-lit, and those illuminated from above or below, as distinct from those which require no lighting, or simply a bright background.

With dark or only slightly transparent crystalline precipitates (e.g., Berlin blue), a bright background works best. With transparent crystalline materials, well-directed side-lighting calls forth the best results. Partially transparent hydro-gels are best appreciated in front of a dark background and illuminated from below; this provides a brilliant "dark field" illumination. A difficulty with this is that the precipitate often covers the light source; top-lighting can also be tried, but this is often less pleasing and more difficult to set up. Side-lighting often gives rise to disturbing reflections with round glassware. In certain situations, this can be avoided with rectangular flasks.

Apparatus

A Demonstration Cabinet with Ventilation

It is especially useful if we set up a demonstration table with an opening through which a strong airstream can be sucked away. (Note: this is not very common, and your architects will have to be convinced of the real value of this.) There are many different ways for using such a device. If we want to do experiments on a reasonably large scale, where unpleasant or poisionous vapors are produced, a rectangular box or case with a zinc top and no bottom can be useful (75 cm wide, 60 cm deep, 70 cm high). The bottom half of the rear side is a sliding glass door which can be fixed open, closed or partway open. Because usually vapors rise and the warm air accumulates at the top, it is important to have a chimney vent (about 12 cm diameter) connected to a ventilation opening on the roof, that reaches about 5-10 cm down from the roof.

An important point is keeping the glass walls really clean; old newsprint can be used for this, or spirits can be used. If there are heavy, greasy gasses sinking down to the bottom of the box, we can pretty much close up the sliding door, and the gasses will spiral up and be sucked away. In that case, it would be best to have the air vented from the bottom and fresh air enter at the top. Therefore, a moveable door opening higher up is not convenient, because the rising vapors can escape too easily.

As background, dark or illuminated, plywood can be used, with one side covered with white paper and the other with black. This can be stood vertically near the rear, using two long wood pieces to form a groove. We could also use frames covered with paper or frosted glass that can be illuminated from behind.

Bottom Lighting

For this we recommend a common filament light bulb, suitably strong (100-150 watts). The bulb should not form a beam, but should be as diffuse as possible. For example, we could set a beaker or cylinder on a tripod, without the wire gauze, and set the lamp underneath. A paper cylinder should surround the bulb and extend up almost to the tripod circle; the inside of the cylinder should be black to prevent the light from reflecting out sideways. White paper primarily prevents too strong a heating, and it additionally concentrates the light more. On the one hand, we must provide good ventilation for cooling, and on the other hand take care that as little light escapes sideways.

Water-filled flasks will appear very impressive with such lighting in a dark room. They will seem nearly invisible to the eye, except for the surface of the liquid and the edges of the glass vessel glowing strongly. The slightest imperfections in the glass walls, such as air bubbles or fingerprints will show immediately. The glassware should be totally clean. Under these conditions, most of the light is kept inside the vessel by total reflection and is directed up through the water. It is really necessary to protect the light bulb from accidental drops of water with a glass sheet; it is tricky to prevent the glass sheet from cracking if it gets too hot, however.

Of course it is easy to make a better light box. A cardboard cylinder, lined with aluminum foil and painted black near the top is mounted on a plywood base; on top, we attach another plywood sheet with a opening corresponding to the cylinder; we cover it with a glass window, maybe hexagonal with the edges protected by thick tape. Ventilation holes are punched in the cylinder at the top and bottom, but must be small enough not to let light out. The glass beaker stands on nested wooden rings, stacked to make the opening larger or smaller.

Examples of Experiments

1. Set an empty glass cylinder on the light box; pour a thin stream of water into it. The air bubbles caught up with the water form a wonderful silvery sheen.

2. A 3-liter beaker is filled with tap water and set up on the light box. Slowly add saturated copper sulfate solution, a spoonful at a time. Initially, the water will have a light blue color, but eventually a weak blue turbidity (opalescence) will be visible with strong lighting (hydrolysis of the salt). Now, add a spoonful of hydrochloric acid; the turbidity is consumed and a clear solution remains. We can help this by gently making little vortices with a glass stirring rod; then we see amazing forms emerging in the turbidity, which remind us of cirrus clouds. Now, add some sodium hydroxide solution; wonderfully luminous, delicate flakes of copper hydroxide emerge. By adding hydrochloric acid again, these can be dissolved again. Then, carefully add ammonia (ammonium hydroxide) dropwise to the solution. Deep, blue colored vortices are formed surrounded by a pale blue precipitate. The vortices rise to the upper part of the beaker, forming a deep blue but transparent layer

there. Again, we can add sodium hydroxide; copper hydroxide is formed, even more beautiful than before. Now, carefully add some concentrated sulfuric acid. This sinks immediately to the bottom, causing some sloshing motion, and dissolves part of the precipitate, while the rest remains floating above. We can produce an enchanting image in this way, consisting of various layers. If we allow it to stand for a while, there will be slow changes which will often go through exceptionally beautiful stages.

This experiment shows clearly how we must carefully account for the various densities, but how we also can use them to achieve specific results.

3. Place a 3-liter Erlenmeyer flask on the light box and fill with a slightly acidified solution of copper sulfate. Suspend a yellow crystal of potassium ferrocyanide (K_4FeCN_6) at the top (stick it on a wire loop with lip-balm or beeswax); a red-brown precipitate forms and sinks as luminous flakes to the bottom of the flask.

If we use iron sulfate instead of copper sulfate, the result is dark blue flaky precipitate in a red-brown solution. This phenomenon is most clearly seen against a bright background, while the former copper sulfate version is best shown bottom-lit with the light box.

In general, we often get surprising images in our experiments, if we do not mix two substances which react by using solutions, but instead introduce one as a solid into a solution of the other.

Other examples: copper sulfate crystal in a solution of sodium hydroxide lye; a piece of soda lye in a solution of hydrochloric acid.

4. The light box: With the illuminating apparatus, one can greatly enhance the visibility of smoke or haze (mist). As a beautiful example, one can show the formation of ammonium chloride by bringing together hydrochloric acid gas and ammonia gas (ammonium hydroxide reagent solution). We place a large Erlenmeyer flask or wide glass cylinder on the light box. Pour some ammonia in the flask and place a filter paper over the opening. On this, we pour a few drops of hydrochloric acid (concentrated reagent). A brilliant white smoke results that sinks to the bottom in twisting threads or streamers. If we do the experiment in the reverse order, with hydrochloric acid at the bottom and ammonia at the top, then it develops much more slowly.

As with every other chapter of this book, this chapter assumes that the teacher has mastered the usual scientific knowledge and laboratory skills to a certain degree, before taking up teaching. So, again, we will not strive here for a complete explanation of all details.

Of course, we will also use the usual methods in teaching. There are excellent books in this area. I wanted primarily to enrich the possibilities for doing demonstrations and experiments and further to find new ways to allow certain aspects of matter, which often remain hidden, to reveal themselves in the experimental scene. When one has used this image to school oneself, then the eye will, as it were, penetrate deeper into the phenomena, meet the phenomena; by developing the experimental skill adequately, the hidden is drawn out (unlocked), and the phenomena come forth to meet the perceptive eye.

III

Salts, Acids and Bases: Chemistry in Grade 10

Crystallization of Salts

In this class, we start with salts—that is with substances which being crystalline are totally earthy, but earth that is strongly, regularly-formed and transparent.

With that we have also exactly struck a theme which is so important for students of this age. They have gone through puberty, and the subtle connections they still had with the cosmos have been severed now, for the most part. They have now become earth-dwellers and must learn to orient themselves with clear thoughts in this realm, which is for them as yet, unknown. Their life, which for their consciousness takes place in first instance on earth, must be given structure through the power of thinking. Their thinking must be so strengthened that it can order the multitude of individual phenomena, but it should also remain so sensitive, so subtle, that it can really grasp the noble organization of the universe.

The whole main lesson block should have the following style: a pronounced clear conceptual structure, that orders the wealth of phenomena, through which it becomes possible to understand and even to use this structure. A nobly shaped, beautifully colored salt crystal can actually become an ideal for us, when thinking about this main lesson.

We start by showing a number of crystals. From the beginning, it is a good habit to use certain devices, so that the impressions become as expressive and characteristic as possible. Such a transparent substance with its shining surfaces doesn't show up very well on white paper. On black paper and with illumination from the side it shows an exiting beauty.

We could perhaps let some salts crystallize beforehand: table salt, copper sulfate, potassium nitrate and others. A wonderful fresh brilliance and beauty meets the eye, if we choose the proper lighting.

If now we reflect on the essence of salts, then we should stress their marvelous precise shape, their transparency and clarity. The real wonder of salts only becomes evident when we consider them in relation to the whole of nature. The salt is earthy, dense, rigid matter, and yet it opens itself to a large extent to the play of light. Actually, the earthy region beneath us is dark, dull, silent and pressing, while in the region above the earth light weaves in a very rarefied, clear world. The greatest contrast exists between the realm of light above and the realm of matter below us. The salt crystal, which actually belongs to the realm of heaviness, is nevertheless largely open to raying-in from the realm of light.

We may take the following thought for our own consideration: In primeval times, matter was separated from the realm of light and condensed. Light has become rarefied; matter has lost its ability to shine on its own and is subject to gravity. Such a crystal shows its isolation from light to a high degree. In its essence it exhibits the *memory* of light, but just because of that it can make such a dull, empty impression on us. If we illuminate it in such a way that it seems as if it is shining of its own accord (backlit or from beneath), then we can achieve a splendid effect. In a sense, we raise it up from its fall and bring it back to its origin. That can be done in various ways. For example, cut small holes in a piece of black paper; in front of every opening place a well-formed salt crystal. Then, darken the room and illuminate the holes one after the other. Then the crystals will radiate beautifully. We can also show crystallization in front of a dark screen strongly illuminated from the side. For example, take a saturated solution of table salt and add some concentrated hydrochloric acid. That gives rise to a magnificent brilliance of crystals, which very often may be so small that individual ones cannot clearly be distinguished. When the crystals are somewhat larger, they are seen forming themselves like delicate clouds, slowly sinking to the bottom, white and pure as freshly-fallen snow.

To experiment on a large scale, use a 3-litre conical flask filled with a hot, nearly saturated solution of potassium nitrate. When the carefully wrapped up flask is allowed to cool slowly, the crystals build enchanting forms—impressive highly transparent shapes with a subtle play of color (the slower, the better the crystals—editor's note).

Crystallization in general, but especially with table salt, evokes the image of the primordial coming-into-being of matter, the original densification of spirit to matter. A sort of "fall" takes place, an expulsion from a surrounding world, but in the magical purity of the forms, a memory of its origin remains.

It's a good thing to let the students totally experience how, in the liquid state everything is in flowing, swirling motion, and how crystallization is something like a protest against fluidity repeated again. Time and again some salt will shoot out of the liquid and becoming rigid and sharply formed, enclosed by planes; repeatedly the fluid movement is thwarted. This is especially apparent with cubic crystals such as table salt. Such a crystallization consists of many jumps from one stage to another that are specially dramatic. Every time, there is a transition that seems impossible but nevertheless happens.

Now is the time to investigate the salts in our body. We find them above all in the skeleton. Here this also happens through a continuous deposition and fixation of salty substance from the flowing blood. However, it is remarkable that here the salts don't take on their own form, but totally adapt to the necessities of the human body.

If the teacher feels justified in mentioning it, then they can draw attention to the fact that especially in relation to the crystallization of table salt, we have here something of a process that also occurs very subtly in our whole organism when we form a thought. Just as in a crystal the earth opens itself to the light, so in ourselves matter must, so to speak, take on crystalline form, if our bodily organs are to become transparent for a spiritual reality, to which our thoughts give expression.

Phenomena Occurring when Salts Dissolve

Once we have occupied ourselves with these considerations for a few days, we can move on to the process of dissolving of salts. We place an orange-colored potassium dichromate crystal, for example, in a beaker glass with water, and let the students observe what happens. The crystal slowly gets smaller and takes on a rounded form, while all around the crystal the water begins to share its color. If someone has enough courage to taste the water, one can taste its saltiness (HAZARD with potassium dichromate—try it with dissolving spears of saltpeter). The crystalline salt, first enclosed within its walls, lets go of itself and loses itself into the surroundings. This process is like a longing for infinity, for perpetual expansion and rarefaction.

Crystallization is a continuous falling-out of unity into multiplicity, a sort of splitting-up, but at the same time a drawing into specific, individualized forms. From solutions of various salts we see each salt taking on its own shape. During dissolving the opposite takes place—the multitude of salt crystals become one with the water. Many salts even dissolve without interfering with each other.

If we suspend potassium permanganate crystals just under the surface in a large glass cylinder filled with water, a particularly lovely process of dissolving can be observed. For example, we can attach the crystals with Vaseline to the underside of a floating cork or to a loop of copper wire. The strongly-colored violet solution sinks, and thereby creates a number of elegant streamers with eddies and vortices. The best results are achieved against a white background, which is strongly side-lit. Then it is possible to observe it all even from a distance.

To be able to do calculations later on, it is now time for a preliminary discussion of various concepts, which will be developed more precisely later on: solubility, concentration, saturated, unsaturated, supersaturated solution, water of crystallization, osmosis, diffusion, heat of solution and of crystallization, elevation of the boiling point, depression of the freezing point. We will treat these in another chapter.[1] Here it is necessary only to point to the fact, that all these concepts can be related to an inner experience of the specific dynamics of the phenomena, and only later be developed into mathematical relationships. If we don't proceed very carefully here, we only add to an aspect that is already being cultivated to a high degree: forming concepts which are very precise and make it possible to manipulate natural phenomena, but are not very helpful in creating a living picture of reality. During the last centuries man has ever more moved away from nature by using such concepts, and has become more and more a prisoner of his own thoughts. We are concerned, therefore, with preserving just that quality that is specific to children, namely their unity with the surrounding world. We must teach the students to develop thoughts which do not break their connection with the world, but elevate them out of the sphere of mere instinctive feeling, to a sphere where conscious precision rules.

It is, of course, good and necessary to bear in mind all the phenomena that occur in life, specially those that occur in human beings. A few indications may point the direction in which they may be found.

We have already discussed the relation between crystallization with thinking, and with the formation of the skeleton. We can also show how aging is accompanied by a continuous deposition of salt in the skeleton, and finally also in the walls of the arteries.

During sleep a large amount dissolves which had been deposited during waking. During the growth of the skeleton a continuous dissolving of certain parts takes place, while at other places new substance is precipitated.

We can speak about the blood as a salt solution, and about the kidneys as strict regulators of the salt levels. This constant salt level is most closely connected to the fact that the tissues, and in particular the red blood cells, depend on a specific osmotic pressure. That must be taken into consideration when giving a blood transfusion after a heavy loss of blood. We can discuss the problems of osmosis that a young salmon encounters when going from fresh to salt water, and vice versa for the adult going to spawn. We can describe how a frog doesn't drink, but a nearly dehydrated frog only needs to place its finger in water to fully recover.

So, there are many topics, some of which are important or must be dealt with at any rate, while others may be included as desired.

Splitting of Salt by Fire: Acid and Alkali

When we have dealt extensively in this manner with these phenomena relating to salts, especially those that occur in interaction with water, we can examine how fire acts on salts. For example, a piece of blue copper sulfate is held in a flame. We see that it gradually loses its transparency, and it becomes opaque white, then brown, and even black. It becomes an inert, earthly, insoluble mass. If we heat the crystal in a heat resistant test tube, then we notice how water of crystallization initially appears as vapor. That is the reason for the cloudy white substance. Subsequently, an acidic pungent gas evolves. This gas colors blue litmus paper red.

If we heat a salt like calcium nitrate, then brown, acidic fumes are generated, and we are left with a moderately soluble milky white substance, which is chemically reactive. This substance colors red litmus paper blue. So we have before us the process of splitting salt. Out of a single substance, which for our perception formed a complete unity, entirely different substances have come about. The original substance has disappeared to the same extent as new substance is formed. These are typical examples for the decomposition of salts. Now we must direct our attention to the transition, to the "interval" between the first state of the substance and the next, as we already did with crystallization. However, we will confine ourselves to the essentials, in that we will not focus on the water of crystallization. We observe, therefore, a well formed, transparent substance changed to a shapeless mass. The "memory of light" spoken of earlier, disappears, and in its place something arises that makes an earthly impression on us.

If we turn our attention specifically to the "phase" of vapor production, then we observe how a gaseous, fine dispersing substance is freed from a clearly defined, rigid, heavy substance. The immobile salt changes into a mobile, active, here even aggressive, gas.

First, there was a piece of earth that was transparent (open to light); now, on the one hand, we have a piece of earth that shuts itself off from light (is opaque), and on the other hand, a substance that seeks out the light filled space and identifies itself with it. The substance that remains behind has retained the solidity of the original salt. Therefore, it is called "the basis" or "base" of the salt. The substance which escaped has taken with it the transparency.

We already mentioned previously, that it can be economical to make a distinction between topics which are dealt with in more detail, and others where we don't do more than give a list of facts. We can use this principle now.

After discussing the creation of acidic and alkaline substances, due to the decomposition of salts, with great care we can list the names of the most important acids and bases:

Some important bases	Some important acids
sodium hydroxide ("caustic soda") potassium hydroxide ("caustic potash") calcium hydroxide ("slaked lime") magnesium oxide ("magnesia") ammonium hydroxide ("ammonia") iron oxide ("iron rust") copper oxide ("copper rust"[2])	sulfuric acid sulphurous acids nitric acid nitrous acid hydrochloric acid silicic acid phosphoric acid phosphorus acid carbonic acid

These substances can also be exhibited, at the same time and, if one has sufficient time, also be characterized with the aid of typical phenomena.

Now we consider the acidic and the alkaline state in greater detail. For example, place three beaker glasses filled with water colored with litmus solution. In the left beaker we pour some sodium hydroxide solution, so that the liquid colors blue; in the right beaker glass we pour hydrochloric acid, so that the liquid turns red. Then we add to the left beaker glass some hydrochloric acid, to the right one some sodium hydroxide solution, whereby the colors change.

It can be made clear to the students that red is a typical expression of the active, aggressive acid, blue that of the more passive base. Although there are other indicator substances that reveal the acidity by a color, and that these colors can be quite different, is not so important here; most are synthetic products which have very little relation with life in nature. Instead, litmus shows us the original and typical phenomena, which we have isolated from nature and are demonstrating on the laboratory bench.

Although litmus is obtained from lichens, it does have the character of a flower pigment. Now, draw attention to the variety of color changes in flowers, for example in forget-me-nots when they change from red to blue. Also, the extract of red cabbage could also be shown as a substitute for litmus.

We arrange three beakers: one with dilute sodium hydroxide, one with dilute hydrochloric acid, and one with tap water; some litmus solution can be added to all three beakers if desired. Then, we have the whole class come up and let them feel how the first solution makes the finger slippery, the second counteracts that, making them "squeaky" or "rough." The third glass is used to rinse the fingers. Remind the students of the similar slippery feeling they had when dipping their fingers into a solution of soap or washing-soda (sodium sulfate).

We now try to discover where we can find acid and base in our bodies. Our whole skin is slightly acidic because of perspiring. Our blood in contrast has to be slightly alkaline. Our saliva is somewhat alkaline, our stomach quite acidic, the intestines again alkaline. In general we can say, everything that is directed outwards (open to the outside world) is acidic, and everything that is directed inwardly is alkaline. The function of the stomach, besides being a storage place, is in the first place to defend against harmful influences that arise from assimilating food. Hardly any nutrients pass through the stomach wall into the blood. A muscle that contracts tends towards the acidic; on relaxation the alkaline comes to the fore.

In 1929 a little book ***Ten Years of the Free Waldorf School*** was published .[3] It contains a very inspiring article by Eugen Kolisko (1893 - 1939) entitled "On the development and forming of science curriculum in the Waldorf schools." He quotes therein a remark of Rudolf Steiner made in a class where they were just discussing the making of acids and bases: "All right, now you have seen all this demonstrated. Now also think, does any of that take place in my body? When you move your limbs, some acid is always formed; but when you are really quiet and are only active in your head, then something alkaline-like forms in your brain." Such a remark is at least as important for the teacher as for the students. From it we can gather in which direction to look.

In the ***Curriculum Indications***,[4] Rudolf Steiner particularly emphasizes the circumstances in the bees, the polarity between the acidic nectar and the alkaline bee blood.

In this context, the contrast between acidic and alkaline types of soil is also very interesting.

Salt Formation by Adding an Acid and a Base

Now we will turn to the action of acids and bases on each other. We take, for example, a beaker glass full with water and add copper oxide, whereby the clear water becomes dark and cloudy. (The copper oxide can be prepared beforehand from a hot solution of copper sulfate and a hot solution of sodium hydroxide. The freshly prepared copper oxide reacts more quickly). Now add a dilute solution of sulfuric acid. We can notice that the dark mass clears and changes into a crystal-clear blue liquid. Also, if water is heated beforehand, everything then proceeds much more quickly. Illuminating the moderately large beaker glass from underneath makes the process visible from a large distance, and many exciting details become observable.

This experiment shows us again a certain peculiarity of acids. They have the ability to consume excessive earthiness and to open the way for light. Precisely because of the illumination, this effect of the acid comes so clearly to the fore. If the blue solution were then evaporated, the well formed, deep blue, transparent copper sulfate crystals would be obtained. A process just as dramatic as the decomposition of a salt has taken place. Here, a shapeless and lusterless mass changes into clear crystals. Also, an active, mobile acid changes into an inactive, motionless salt.

We can now pour some concentrated hydrochloric acid into a test tube and bring a piece of sodium hydroxide in contact with the surface. A sharp sizzling can be heard, while a number of clear salt crystals fly away like dust and sink like snow to the bottom. The experiment can also be arranged in such a way that it takes on a slight sensational touch. Place the test tube with hydrochloric acid in a test tube stand, and drop a pellet of sodium hydroxide in the tube. The liquid begins to boil violently, and the tube begins to dance up and down.

When different ways of salt formation have been dealt with extensively in this way, then a lot of scientific knowledge can be introduced in a concise way.

We discuss the rule according to which we can, in principle, combine every acid with every base and always arrive at a salt. This can be done diagrammatically as follows:

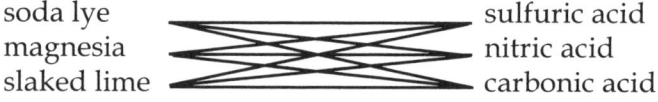

```
soda lye                          sulfuric acid
magnesia                          nitric acid
slaked lime                       carbonic acid
```

Then we discuss how we find the names of the salts and make a list of the acids and their related salts. Then we make a list with their common everyday names and also their scientific names.

Transition to Chemical Equations

Now show how the general rules for the formation and decomposition of salts can be put into simple schema:

salt decomposition: salt ⟨ acid
                         ⟩ base

or, even more abstractly:

$$\text{salt} \longrightarrow \text{base} + \text{acid}$$

salt formation: acid + base ⟶ salt + water

With the help of these general schemes, specific reactions can be inferred by inserting the names as required. So for example:

caustic soda + sulfuric acid ⟶ sodium sulfate + water

slaked lime + carbonic acid ⟶ calcium carbonate + water

Or a salt decomposition:

calcium carbonate ⟶ unslaked lime + carbonic acid

In such a way the foundation is laid step by step for the subsequent use of reaction equations and formulas.

It isn't really possible to introduce formulas on the basis of the curriculum of grade 10. If the path going from the whole to the parts is consequently followed, if the whole is taken as the origin, and the parts that come from that as secondary, we just don't get that far. Besides, it isn't yet possible to treat the chemical elements so extensively and in such a lively way that could counterbalance such a far reaching abstraction. If we do it anyhow, a distinction is very easily created between the students. The more intellectually inclined will fall into the illusion that they now really have understood it, while in reality they have lost the connection with existence, and besides they don't grasp the difficulty of the formulas (they are abstractions). The formulas become an intellectual game. For the intellectually less-inclined, the formulas become a more or less unfathomable and unconquerable wall that discourages them from taking a real interest in the world of matter.

Yet, something important must be added. We have mentioned at the beginning of the discussion of grade 10 that we must be careful that the conceptual development is clear and lucid. The general reaction equations as mentioned for the salt formation and salt decomposition are in agreement with this; the area of the formulas, however, contains so many facts that the clarity quickly diminishes.

We will see that the full treatment of formulas, if postponed until grade 11, are better discussed from a different angle and then present an unexpected possibility for schooling.[5]

Besides, formulas have the disadvantage of unintentionally diverting our attention from the fact that every substance, be it an element or a compound, presents itself as a wholeness. Talking about sulfuric acid is in harmony with this fact. Writing down the formula H_2SO_4 however, specifically draws attention to the fact that this substance can be produced in a reaction from other substances.

However, it is possible to abbreviate our "general" equations somewhat and make them more lucid. We can do that in the following way:

$$\text{base} + \text{acid} \rightarrow \text{salt} + \text{water}$$

$$B + A \rightarrow BA + \text{water}$$

$$\text{salt} \rightarrow \text{acid} + \text{base}$$

$$BA \rightarrow A + B$$

Strictly speaking, the formula BA for a salt is already too analytical; in the end every abstraction, even the most circumspect, removes us from reality. But doubtless we have to put up with this.

On getting this far in the treatment of the curriculum, we have reached an important point. The students now possess the concepts not only to be able to gain insight into the processes to a certain extent, but also to know beforehand how a chemical process will proceed. They can even put reactions together for a specific purpose. For example, if we want to make copper nitrate, then we have to react copper hydroxide (copper base) with nitric acid. In this domain, students can predict and then check through experimentation if their thinking was correct.

Weight Relationships

If you wish to introduce the concepts of atomic and molecular weight later on, then I think it is appropriate to point out here that all these reactions proceed according to definite weight proportions. It is easier to treat this fact now as an expression of a strict harmony in nature than it is later on, as long as you don't yet have to do quick calculations. It is quite possible to proceed from the equivalent (combining) weights, e.g.:

caustic soda	40
sulfuric acid	49
sodium sulfate	71
water	9

caustic soda + sulfuric acid → sodium sulfate + water

40　　　　　　49　　　　　　　71　　　　　　2 x 9

It seems to me absolutely necessary to remain in the realm of abstract numbers, because only then will it be clear that we are dealing with ratios (of weights) and not with absolute weights. Nevertheless, it is possible to make simple calculations with these simple numbers. For example, we can find how many grams of sodium sulfate we get, when using 80 grams, or else using 2 grams of caustic soda (sodium hydroxide).

$$\text{caustic soda} + \text{sulfuric acid} \rightarrow \text{sodium sulfate} + \text{water}$$

$$40 : 49 = 71 : 2 \times 9$$
$$80 : 2 \times 49 = 2 \times 71 : 2 \times 18$$

or (dividing all weights by 40):

$$2 : 49/20 = 71/20 : 18/20$$

If we give the students a list of the equivalent weights, then any number of calculations can be done.

Oxygen, Oxidation, Burning, and Rusting

We now demonstrate that acids come about when we burn certain non-metallic substances, and bases when we let metals rust. So, we must investigate in more detail the role oxygen plays when an acid or a base is formed.

It is quite possible to start historically and discuss how oxygen was discovered by Priestly. This presents us with the great advantage of being able to follow the historical development, which is mostly accompanied by an instructive struggle for the clarification of concepts. Moreover, this gives the opportunity to include biographies. This is particularly important for the girls, who more than the boys are inclined to assimilate everything with the soul and relate to people.

Priestley produced an unknown gas by heating mercury "rust" (oxide) and established that this gas stimulates the breathing process and also keeps the burning process going. It was, therefore, a gas with "enhanced air qualities."

Now, for example, burn some sulphur or phosphorus and demonstrate how the rising vapor and smoke will color a piece of moist blue litmus paper red. For completeness sake we can demonstrate that these substances will only burn if there is sufficient oxygen in the surroundings. If necessary, we can enhance the burning process by supplying pure oxygen (although this is not usually necessary, except with charcoal).

We then place a piece of iron in humid surroundings and let it rust. The tough, shiny metal slowly tarnishes, becoming rough and finally becoming a brittle, reddish-brown, crusty mass. The most interesting thing about this process is its slowness and lethargy—but this will not usually mean much to the students. Best of all, also, is to show the students a piece of severely rusted iron.

Going further, a bright piece of iron can be heated, and then copper. You can observe then how quickly the shiny surfaces tarnish. You can then discuss, and perhaps demonstrate, that the resulting copper oxide is the same substance as the base which we got from the decomposition (heating) of copper sulfate salt.

We now compare these processes of burning and rusting. In principle, both are caused by the action of oxygen. If both are expressed in chemical equations, then they both look quite similar.

$$\text{sulphur} + \text{oxygen} \rightarrow \text{sulphur oxide}$$

$$\text{copper} + \text{oxygen} \rightarrow \text{copper rust}$$

However, looking at the dynamics of the process and the qualitative image they give us, each processes shows a totally opposite character.

Burning proceeds quickly. It is a particularly striking and radiant process, because of the production of light and heat, and through the pungent smells that normally occur also. The substance that was before us initially as a solid object becomes totally consumed and changed into a mobile, space-filling gas. The whole process is primarily directed upwards.

Rusting is a slow process. The smooth, bright surface of the metal turns itself, as it were, away from the light to become tarnished and earth-like. Very little heat production can be noticed. While the burning substance goes with the oxygen into its sphere of gases, the rusting substance draws the oxygen down into the earthy realm.

Here is something mysterious, in that combustion only occurs when a certain (kindling) temperature is reached, while rusting also takes place even at low temperatures. However, this riddle can form the key to understanding the pattern that underlies all this. Combustion occurs through heat, rusting through moisture. During combustion mostly gasses and other volatile substances are formed. During rusting earthy substances are formed.

Combustion:	caused by heat	Firey
	result: gas/vapor	Airy
Rusting:	caused by moisture	Watery
	result: earthy substance	Earthy

We observe that the relatedness of both processes becomes clear, as soon as the four elemental qualities are taken into consideration. From that viewpoint it is particularly interesting to note that metals can be protected from rusting (watery) using an oily coating—stopped by combustible, "Firey" substances, while we fight fire with water.

Now we demonstrate how the products of combustion of substances like carbon, sulphur and phosphorus become acids in water, while metal oxides appear as bases. All this contributes to clarifying the typical properties of acids and bases. The acid, as it were, is the chemical herald of the two higher elements (Firey and Airy). This explains a lot of their aggressive mobility and their ability to open up the way for light. The alkaline is the chemical herald of both lower elements (Watery and Earthy). That explains their property of avidly absorbing acid vapors and binding them. Strong bases like caustic soda have more affinity to the Watery element; the insoluble bases, like the rusts of the heavy metals, have more affinity to the element of Earth.

Later on we will see how the properties of silicic acid are generally opposite to those mentioned, for silicic acid contributes a lot to the solidity of the earth, and it also has a special relationship with the cosmic aspect of nature. With bases, ammonia forms the exception; it is a volatile gas instead of a solid.

The Importance of the Approach Chosen

We could ask again, why does all this have to be treated in such a complicated way? Why don't we just start the block deriving acids from combustion and bases from rusting, and then the salts from reacting the acids and bases? From our description built up so far, we can make various answers. If we had started with burning and rusting, this would have followed an analytical method (emphasizing weights as Lavoisier did), even though we are seemingly proceeding on a synthetic path. But, this way gets into treating the substances that constitute a higher unity (such as acids, bases, and salts) as if they were made up out of these parts. Unconsciously, we lose sight of the wholeness. Taking the opposite approach, as we did, then we keep the wholeness in sight, even when going into the greatest detail.

From this it follows that this method is in harmony with the laws of evolution of a living being. This always begins as a unity with slight variations and goes on into ever finer differentiated detail. Using our method means we are working in accordance with the development of the child, which of course also follows the general laws of life.

Through the analytical methods widely used in science, there is always a danger of destroying the image of nature, resulting in unintelligible parts. Of course, this forms very well-defined concepts in relation to the phenomena, but then we wouldn't be very concerned whether they were an expression of the *context of reality*, within which the phenomena are embedded.[6]

Indeed, with our exact concepts, and infallible definitions we are continuously cutting parts out of the whole of nature. We are often so impressed by the clarity of the concepts, that we start to view the details as independent entities. The creators of mechanical theories, for example, suffer from this malady. They want to explain reality by using such parts from which they can develop the most lucid thoughts. If, however, you start from the whole, as we have been trying to do, then the concepts of the details developed will always remain in a harmonious relation with one another and with the wholeness of nature. They then acquire an aspect which reminds us of the organs of a living being. In this way, we create in the students quite a different orientation towards life. They develop the habit of keeping sight of the whole of nature and come to their own judgement and point of view from that whole.

In connection with the formation of acids and bases, it can be very fruitful to discuss the great French scientist Lavoisier. He is indeed the actual discover of the role of oxygen in this process. But to clarify the direction of his work, it is really necessary to talk about the alchemists. We can explain how the serious alchemists weren't after riches, but rather their striving was directed towards inner development with the help of chemical experiments. Related to that is also the strange names they gave, which is a way of expressing processes in imagistic form. Here we are dealing, on the one hand, with suggestions and hints about substances and their states, but also, on the other hand, with descriptions of inner changes occurring in human observers. During experiments they concentrated specially on the image-like aspects of the processes, on which we also have focused. They perceived the divine in nature in these images and the changes in themselves.

In Lavoisier's time the alchemical movement was long past its prime; it had become confused and decadent. Indeed, he has contributed a lot to the thorough extinction of the residue of alchemical views.

For him, the balance was the most important instrument, and this means that he tried to approach all phenomena from the aspect of weight. By that approach, the pictorial was totally eliminated. It is very remarkable that he was able to weigh oxygen before he even knew that it existed. In a closed vessel he let zinc oxidize and was able to show that something went from the air into the rusting metal. The total weight of the vessel and contents remained the same; however, the metal increased in weight. When he heard from Priestley about his studies of combustion in an oxygen atmosphere, he quickly developed his theory of combustion.

He also contributed a lot to the modern non-pictorial nomenclature. He proposed the name "oxygen" (from Greek, oxoos-gen = acid producer), and also the name "oxide" for an oxygen compound. He is also associated with the Law of Conservation of Matter. As a practical person he was primarily interested in an experimental method. He based his experiments on the conviction that the total weight of the substances taking part in a reaction remained unchanged. That really is only a working hypothesis. He most probably never would have drawn the conclusion, as materialists later did, that everlasting existence, and not even the human soul, had no divine being as its essence, but only matter. If matter can't be created or destroyed, then Creation isn't possible. Because of such considerations, in a sense God was robbed of his power, and matter was set upon God's throne.

Lavoisier is seen as the founder of modern chemistry, and in practice, chemistry views the material world primarily from the viewpoint of matter and energy. Maybe it is right to discuss such things with the students, mostly because we've learned since then that matter can actually be destroyed and created. Materialism, on this level, has been disproven. Besides, we can show the one-sidedness of such considerations precisely by studying rusting and combustion. If only the weight is considered, then the qualitative differences between these processes disappear. If, however, we pay attention to the image and the process, then the greatest differences can be discovered. Up to a certain extent it makes sense to consider rusting only from the aspect of weight. However, fire is in constant contradiction with gravity, and in the same manner, it is in harmony with the radiant light side of nature, just as rusting is related to gravity. With such thoughts the students are not only given a harmonious world view, but it should even be possible to help them find their own relation to nature. The young students have just recently deeply experienced in puberty the falling apart of human beings into male and female. At the same time, they became caught up in an interplay of soul forces that, although they may elate or fill them with idealism, can also pull them down. They feel fractured.

An image of this drama stands objectively before them in the decomposition of salt into acid and base, in the process of combustion and rusting, and what may be followed up from there. While they feel themselves more or less ejected from the care of a divine world-unity, this image may also contribute to the feeling that what they are going through now is also founded on profound laws of the universe. This drama becomes part of the world-drama. Their task is to acquire a new dignified bearing in the interplay of forces in which they find themselves. Once, they had to achieve an upright stance, between the world of light above and the world of gravity below them. At this age, they become mentally aware that there is something which wants to pull them into illusory worlds and also something that wants to pull them down and bind them to unconscious forces. They must learn to stand spiritually upright in this mental field, as previously they had to do this with their body. They must consciously find a new spiritual equilibrium. The curriculum, as reviewed above, may be a support for this.

Reduction

We have now reached a point where we can discuss the phenomena of reduction more closely. We place some brown, powdery iron oxide in a heat resistant glass tube. We then conduct natural gas (methane) or hydrogen gas through the reaction tube, and heat it where the iron oxide is. We observe the powder becoming gray-black, while a precipitate of water becomes visible on the walls. We can put this directly into a reaction equation. First we had iron oxide and hydrogen gas, now iron and water.

$$\text{iron oxide} + \text{hydrogen gas} \rightarrow \text{iron} + \text{water}$$

This can also be expressed as follows:

$$\text{iron oxide} - \text{oxygen} \rightarrow \text{iron}$$

$$\text{hydrogen gas} + \text{oxygen gas} \rightarrow \text{water ('hydrogen oxide')}$$

After it has cooled down, the powder can be attracted by a magnet, a clear proof that metallic iron has been formed. It is also possible, while the tube is still hot, to pass air through. We see then that the powder immediately changes color back into reddish iron oxide.

A very impressive way of showing the alteration between oxidation and reduction is by quietly and strongly heating a piece of copper and letting the flame play over the surface. Inside the sphere of the flame, the metal becomes shiny; where the flame is removed, a rapid oxidation starts, that initially displays itself in a beautiful play of colours but ends in a gray-black zone. The longer the flame is moved about, the more splendid the colours shine out.

The students already have an idea of what occurs during a reduction, because photosynthesis has been discussed extensively in grade 9. We should recognize in it the archetypal phenomenon of a reduction process.

During the discussion of the discovery of oxygen, we also mentioned the decomposition of mercury oxide by heat.

Often, reduction is brought about by really combustible substances, like hydrogen gas, carbon, or carbon monoxide. From this, we may realize why all this occurs. The combustible substances are in a state of tension with regards to their surroundings. They are laden with energy, so to speak, which they radiate primarily as light and heat. In contrast, the oxides have reached a totally relaxed state. If we want to conjure up the initial substances from a combustion product (oxide or metal ore), we achieve this tension again by applying light and heat.

During photosynthesis carbon dioxide is changed to carbon by sunlight. When decomposing mercuric oxide, the heat of the gas flame, or, in the case of Priestley, the warmth of sunlight, was used. During the reduction of iron oxide, hydrogen gas "snatches up" the oxygen, and gives some of its energy to the iron.

It's a good thing to draw attention to the fact that not only combustion, but also oxidation of metals, is accompanied by heat radiation. Mostly we don't notice this, because it goes so slowly. But, for example, if the process is accelerated by scattering iron into a flame as powder, the metal is seen forming glowing sparks—a kind of "meteoric burning."

In this way, we can discuss that with every chemical process an interchange takes between the ponderable and the imponderable. And each time, light or heat is either absorbed or emitted.

This point in the review of reduction can be discussed in such a way that we still keep completely within the sphere of the qualitative, connect up with life processes, and put things into thoughts that remain in the realm of images.

Solution of Metals in Strong Acids

Besides the normal way of salt formation by reacting acid with base, there is also the possibility of producing a salt by dissolving a metal in a strong acid. Many significant changes take place during this process.

For example, pour some dilute sulfuric acid in a test tube and add a small piece of zinc. The zinc is immediately consumed, accompanied by loud sizzling, a very combustible gas evolves, and finally, a clear mass of crystals accumulates at the bottom of the tube.

Once again, the qualitative reaction equation for the process taking place can easily be written out:

$$\text{zinc} + \text{sulfuric acid} \rightarrow \text{zinc sulfate} + \text{hydrogen gas}$$

The metal is changed into salt crystals by the acid.

We are here faced with one of the strangest, most remarkable transformations. We have already previously characterized matter as that part of the universe which has been abandoned by light.[7] With the metals there has arisen a sort of contrast to light. If a piece of metal is held against the light, it appears as a dark opaque mass—indeed, it is completely opaque. Metals are the most reflective materials; they send the light back to the highest degree. However, at the same time this reflective ability points to the fact that they are also attuned to light. That becomes more pronounced the stronger they reflect, silver being most pronounced.

If a metal is dissolved in an acid, then the light repulsive properties are totally conquered. Again, acid opens the way for light.

If salt crystals are allowed to form immediately, as in the experiment described above, then the elasticity of the metal, its malleability, changes into the brittle, hardness of crystal structure.

In this context it is specially interesting that transparent solids are never good conductors for electricity, while the best conductors, the metals, reflect light the strongest. Here the polarity between electricity and light becomes very evident. Where one has access, the other is excluded.[8]

From all these phenomena we may conclude that there are polar states of matter; the metal state and the transparent crystal state form such a contrast.[9]

Electrolysis

Having come so far, it is quite possible to incorporate the first phenomena of electrolysis. Again, this will have to be done in a particular "Grade 10-way" and not anticipate the electricity curriculum of grade 11.

Start with a diluted salt solution (not table salt or other halogen salts) colored by litmus. For electrodes one can initially use carbon rods, and later on certain other metals. The results are as follows:

CATHODE (-) tendency towards	ANODE (+) tendency towards
base formation separation of hydrogen gas phenomena of reduction precipitation of metals	acid formation separation of oxygen gas phenomena of oxidation dissolving of metals

In this way we achieve a "phenomenology of the electrode." We can see how electricity, in its polarity, is related to the contrasts we have already discussed.

We should set aside all theory. In the light of practice and application, it is much more important at this point to give a characterization of the phenomena, rather than abstract explanations.

In this way we even are already laying a certain kind of foundation for the physics block (Electricity) in grade 11.

Finishing the Main Lesson Block

Acid or Base Substitutions—Double Substitutions (Displacements)

We now come to a part of the main lesson block wherein many new phenomena occur, which often include the most impressive ones we can observe. Generally, we will be able to use the concepts we have learned up to now. That makes possible constant repetition and practice. The students learn to play, as it were, with the concepts. They must be made to find explanations for many cases and to devise examples which can then be experimentally verified.

Furthermore, it is a question of finding all the possibilities that can occur when combining salts with acids and bases. Chiefly, there are three such combinations: base and acid, acid and salt, salt and base. We bring them together and observe the results.

For example, we begin by adding a base to a salt solution. If we take table salt and copper oxide, nothing will happen, just as with sodium hydroxide (a stronger base) and table salt. But, an impressive phenomena occurs if sodium hydroxide is gently poured into a bright blue solution of copper sulfate that is not too dilute. This results in a magnificent, veil-like translucent blue precipitate. If we follow the process carefully, small globules are observed. Within these globules there is sodium hydroxide; the walls consist of a delicate jelly-like substance, surrounded by copper sulfate solution. These globules, which also can occur with other salts, often look like protozoa. If one of them is destroyed, the contents will indeed flow out. However, a new skin is immediately formed; something like the of healing of a wound comes about. These bubbles are, therefore, rather stable, notwithstanding their delicacy.

Of course, the teacher should try the experiment in all possible variations, to ensure that they can demonstrate the phenomena in the most impressive way possible. Several glass vessels containing different concentrations can be used. If large glass containers and bottom illumination are used, the most enchanting phenomena can be achieved.

In a certain sense, the image becomes more clear if the copper sulfate solution is instead poured into a sodium hydroxide solution. The blue flakes then are quite visible in a colorless solution. Also, something very surprising happens when we hang a copper sulfate crystal in a sodium hydroxide solution—it is possible for pillars of precipitate to form.

Similar experiments can be done, if salts of other heavy metals are used with strong alkaline solutions.

We pose the question to the students: what may have occurred? We can write:

$$\text{copper sulfate} + \text{caustic soda} \rightarrow ?$$
$$BA + B' \rightarrow ?$$

There certainly will be some who will be able to discover the answer. Hardly any other reaction is possible than that a new combination has come about, and in fact this is what has resulted:

$$\text{sodium sulfate} + \text{copper base}$$
$$B'A + B$$

The whole process looks like this:

$$BA + B' \rightarrow B'A + B$$
$$\text{copper sulfate} + \text{caustic soda} \rightarrow \text{sodium sulfate} + \text{copper base}$$

The strong, readily-soluble base has displaced the weak, slightly soluble one. If that is right, then we must be able to dissolve the blue precipitate with a strong acid. Indeed that can be done. Of course, we shouldn't forget to show such an image of dissolving to its best advantage.

If our chain of reasoning was accurate, then the solution must contain sodium sulfate, and the blue precipitate must be some sort of copper base. To prove the first is difficult. However, the students will accept our statement that sodium sulfate really was formed. The second can easily be proven by heating the blue, jelly-like precipitate. Then a brown, black mass of copper oxide is produced. It then isn't very difficult to understand that the blue precipitate is copper hydroxide, a "watery" (hydrated) copper oxide. Such a hydroxide can be viewed as a base that has become stuck halfway to becoming a solid. If we add an acid to a salt, the phenomena can again be quite varied.

On addition of hydrochloric acid to a solution of sodium sulfate, no changes are seen. However, if hydrochloric acid is poured over soda (sodium carbonate), then it starts to sizzle and effervesce, volatile carbon dioxide escapes. Again, it isn't very difficult to understand what has happened here:

sodium carbonate + hydrochloric acid → ? + carbon dioxide
BA + A' ? + A (gas)

From the equation it becomes clear that BA', sodium chloride, must be produced.

sodium carbonate + hydrochloric acid → sodium chloride + carbon dioxide

Again, a general rule is achieved here: if we add a strong acid (HCl) to the salt of a weak acid (carbonate), the weak acid will be displaced and a salt of the strong acid will be formed. If necessary, we can draw attention to the fact that the volatility of the acid plays a role.

For practical purposes, acid displacement is of great importance. It could perhaps make sense to discuss the production of nitric acid, hydrochloric acid, carbon dioxide, and other acids from their salts.

It is important to show with such experiments how during a typical base displacement reaction, a substance falls prey to gravity, and how during an acid displacement reaction, a substance is taken up by levity.

As a last possibility of the interaction between acids, bases and salts we still have that of two salts. If we add a table salt solution to a potassium sulfate solution, no visible reaction occurs. However, using calcium chloride and sodium carbonate produces a white precipitate. It is strange to observe how two, clear fluids, on mixing, produce an opaque, solid mass as if by magic. This precipitate is not jelly-like, but is more likely to be powder-like. What has happened?

calcium chloride + sodium carbonate → calcium carbonate + sodium chloride

BA + B'A' → BA' + B'A

The precipitate could be either calcium carbonate or sodium chloride. Since sodium chloride is soluble, it has to be calcium carbonate. Because of the different combination of acids and bases, two new salts have been produced. The following rule can be deduced: If solutions of two salts are poured together, then a double displacement reaction will occur with the formation of a precipitate, if one of the possible products is an insoluble salt.

Now, which salts are soluble and which are not should be overviewed. In case the students haven't already got such an overview, a list of the most important insoluble salts can be handed out, and also for those acids and bases whose salts are always soluble. The students can then explain and even predict the phenomena.

How double displacement reactions are used as a method of identification of certain salts in solution can be pointed out. With sea water very nice experiments can be done.

The problem of hard water can also be discussed—why soap becomes flaky and why soda softens water.

Lastly, different salts can also be produced via the double displacement reaction, e.g. lead chromate (chrome-yellow), lead carbonate, mercuric chloride or potassium nitrate (as always, CAUTION with toxicity hazards).

Endnotes

[1] See Chapter 6 of this book: "Process of Solution - Phenomena, Concepts, Laws."

[2] In Dutch, "roest" and "roesten" (to rust) are used quite generally for corrosion changes in other metals, similar to when iron rusts.

[3] Special volume of *Zur Paedagogik Rudolf Steiners*, 1929, III, vol. 3/4.

[4] Steiner, R. *Discussions with Teachers* 1919 - 1924, Jan./Nov 1921,17/6, p41; GA 295, Chapter 3.

[5] See "Formula," Chapter IV.

[6] See Chap. VI, the "Concept of Concentration."

[7] See Chapter III, "Crystallization of salts."

[8] Steiner, R, *First Scientific Lecture-course* (Light-course); GA 320.

[9] See Chapter X.

IV

Guidelines for Teaching Chemical Formulas

As soon as we realize that everything we do with students can work both formatively, and also misformatively on their souls, then particular methods and forms of thought which are generally taken for granted become questionable.

For example, learning to use chemical formulas has a profound effect on the notions developed about the nature of substances and about the material processes. Indirectly this affects the observations and, thus, generally the inner life. A far-reaching process of abstraction must have been gone through before we can represent a chemical transformation by a mere formula and reaction equation. This easily leads to ruining the lively interest we should have for the processes of nature. A rich, diverse activity of nature is now represented by a rigid and all too simplified conceptual schema.

If we study Rudolf Steiner's curriculum plan, it is quite evident that *he placed the greatest importance on developing concepts about material substance that are as living as possible.* His indications in the **Faculty Meeting Conferences**[1] indicate that he felt it would have been best not to have to use formulas at all in the classroom. However, in most cases it is probably unavoidable, as formulas are used everywhere else in schools. But this is already a compromise.

To counterbalance this a bit, their use should be guided by well-considered points of view. Then, it isn't even so difficult to find good sides to the introduction of formulas. We can, for example, practice the skill of forming abstractions, which many students find very hard and yet which play such an important role nowadays. Of course, we should pay attention that this process of forming abstractions really proceeds step-by-step. Also, every step should be taken as consciously as possible, and be completely understood before we go further. Every step should be gone through in such a way that the students really grasp it with confidence.

If we really want to follow the curriculum indications, it seems impossible to me to teach formulas in their final form with the accompanying subscripts before grade 11. I have several reasons for this conviction. Firstly, the curriculum in grades 9 and 10 doesn't provide a basis for this. Secondly, students before the age of grade 11 are even less capable of forming a clear concept of the foundations of the system of formulas. Thirdly, the specific thought forms which we discussed as being especially important for grade 10, and which can also be expressed in a certain symbolic way, will be pushed aside or at least become less effective.

The curriculum of grade 10 initially presents us the task of developing a systematic foundation of chemistry as a scientific structure. We have to start with salt, and from that whole, develop acid and base by decomposition. Thus, in the beginning we confront the children with a very typical method of analysis (decomposition) and make sure that they receive a living concept of it. Only after we have brought acid and base together forming a salt can the concept of a "compound" be developed (phenomenologically, as a transformation product, not a combination). Thus, we are following a method that starts by arranging a whole in a series, *a whole, as found in nature*. Putting the parts together again is the next step. With the dissection, the whole still remains as the basis for our concepts and explanations. The analysis, in our conception, doesn't cause the elimination of the whole, but *occurs within* the whole. The whole becomes even fuller and more living through analysis; it spreads itself out as an ordering principle over all the substances that are produced.

From this view, if we now want to introduce formulas at the start, then we would be considering everything only from the point of view of the final decomposition products. And we would sacrifice the whole for focusing on everything that can be observed within it.

Here we have an especially clear instance of a problem that was mentioned in the introduction. In this abstract case we would place our own capacities and method of thinking into the foreground, putting nature into the background and letting her speak too little for herself.

If, for example, we say "$CuSO_4$" and mean "copper sulfate" (or better yet—blue vitriol, the older common name), then that expresses the fact that we are *not* focusing on seeing the well-formed, blue and transparent crystals which lie before us, but rather focusing our attention on the fact that this substance can be put together by mixing three chemical elements in certain weight proportions.

In the discussion of the curriculum of grade 10, I have not only used simple word equations, using the names of the substances, but even introduced a certain type of formula. When a salt is written as "BA," the objection can be made that this whole isn't a whole any more. However, the advantages of clarity are so great, that they override this problematic aspect. Writing an equation in this symbolic way, compared to the conventional form with subscripts, we see that the former symbolic formula affords a much clearer overview.

Taking the base displacement reaction as an example:

$$\text{copper sulfate} + \text{soda lye} \rightarrow \text{sodium sulfate} + \text{copper base}$$

Then the following way of writing can help clarify what is happening through its simplicity:

$$BA + B' \rightarrow B'A + B$$

The conventional equations contain so many facts and thoughts, that initially they can not contribute much to clarifying the issue.

$$CuSO_4 + NaOH \rightarrow Na_2SO_4 + Cu(OH)_2$$

A formula like BA doesn't indicate a particular substance, as is the case with $CuSO_4$, but rather is a mere shorthand indication of a law of nature (bases react with acids).

On the Essence of a Chemical Compound and the Principle of Impenetrability

Textbooks start with the formation of iron sulfide from iron and sulphur. They start with a yellow powder and a shiny metallic powder, which are mixed in a certain weight proportion. Then we heat a part of this mass so it starts to glow; the glow spreads and "eats" through the whole mass. As long as we had a mixture, we could separate both powders quite easily. Both still had their characteristic properties: iron was magnetic, sulphur could easily be melted or even vaporized (sublimated). After they have combined, we obtain an inseparable mass, with totally new properties. All properties of iron and sulphur have disappeared, except that of mass. If some dilute hydrochloric acid is poured over the iron sulfide, then hydrogen sulfide gas is evolved that is recognizable by its awful smell—that is a new property; the fusibility and the magnetic properties have disappeared.

This fact is really appreciated when we realize that in such a process the principle of impenetrability of matter is negated. Initially both substances were *next* to each other and *not in* each other. When they combined, they both entered the same space.

The principle of impenetrability of matter, in my opinion, isn't appreciated enough in conventional chemistry and physics. It is one of the most important characteristics of matter. Matter is defined as "that which fills space." We also find the following definitions of matter: "Now one will nevertheless find, when one proceeds without prejudice, that the most conspicuous attribute of matter is its filling out of space, its extension in space."[2] "Our sense perceptions are related to something extended in space, which we call matter."[3] "The space filling something, which effects our senses, we call matter or substance. A (physical) body is anything of matter which fills a definite region of space."[4]

A conventional phrasing of the principle of impenetrability that closely relates to the previous ones by Rudolf. Steiner is: "Two physical bodies cannot simultaneously occupy the same region of space."[5]

Steiner's correction to this wording is of special importance for our work. "We *should* say: those bodies or beings that are such, that no two bodies *of the same kind* can occupy the same region of space simultaneously, are said to be impenetrable."

In this way Steiner always indicated how already in the fundamental phrasing and definitions of science, a direction can be taken which leads irrevocably to materialism, or leaves the possibility open that there are still higher realms where other laws hold sway.

In general we can say: impenetrability holds for matter; interpenetrability for everything of a spiritual nature. It is characteristic for occurrences in the spiritual realm that it is possible that two different beings can completely merge into one.

What happens during chemical combination is a process in the physical world, which seems to proceed according to spiritual laws. An interplay occurs between physical bodies which otherwise generally takes place between spiritual or soul-beings. We shall see that this insight has far-reaching consequences. Here we have found a hint that should lead us into profound mysteries of creation.

At any rate, there is a great qualitative difference between the occurrences on a mechanical level, where impenetrability plays such a great part, and those in chemistry, where just at the outset, interpenetration of substances in their respective being and essence is most important.

Before we discuss the essence of a "compound" further, we want to show that the groundwork done until now is sufficient to develop a clear insight into the system of formulas. The following method can readily be used in the curriculum of grade 11.

The Foundations of Chemical Formulas

If we are not to be deceived from the beginning, by certain confusions which are quite common, we must first ask: Is the system of formulas, as we now know it, derived from phenomena of nature and, thus, is primarily based on experience? Or, is it based on a conceptual construct, which has been adapted to nature as well as possible? Probably most scientists will answer that the system of formulas is based on experience; however, I am of the opinion that it rests on certain conceptual principles. At any rate, the system can be presented with great clarity if that is assumed to be the case.

We start with purely theoretical substances: A, B, C, etc., which have the additional advantage that we don't have to consider all sorts of specific properties which would distract from the general argument. How will the most simple and clear-cut chemical reaction proceed? A gas A combines with a gas B in the same volume proportions, and a third gas C is produced. Because the "being-penetrated-by-one-another" must also come about in its most simple form, the volume of gas C will equal to the initial volume of one of the participating gases. Only in this case gas A will penetrate the space occupied by B and B that by A.

We can present that as follows:

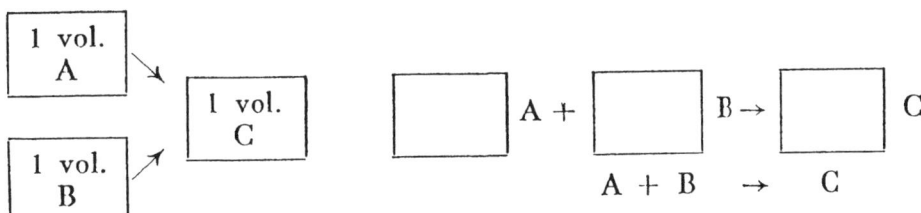

For practical purposes it is easier to write the third substance not as C, but as AB. Then we see immediately which reactants are needed to produce C, and into which substances it can be decomposed.

Our equation then becomes:

$$\boxed{} A + \boxed{} B \rightarrow \boxed{} AB$$

$$A + B \rightarrow AB$$

A somewhat more complicated case is conceivable: a substance A combines with *twice* the volume of B, or in other words, A combines twice with a substance B. If there are no other complications, then the product of this reaction will occupy the same volume as the smaller of the reacting volumes (A), since the repeated combining takes place with this smaller volume.

$$\boxed{} A + \boxed{} B \rightarrow \boxed{} AB$$

$$A + B \rightarrow AB$$

But now, in place of C we cannot write simply AB. We must indicate that 2 volumes of B was used in the reaction. We could choose between the notations A2B, AB2 or AB_2. Initially the choice was AB2; later AB_2 was used.

$$A + 2B \rightarrow AB_2$$

We can imagine cases where A does not react once or twice with B, but more often, although experience has shown that this does not happen more often than seven times.

$$A + 3B \rightarrow AB_3$$

$$A + 7B \rightarrow AB_7$$

So, there are substances conceivable which combine (in the gaseous state) with exactly the *same* volume of another gas. Such substances are said to have "a valence of 1." If A has valence 1, then $A + 2B \rightarrow C$ is not possible. However, $2A + B \rightarrow C$ is possible. A substance which will combine (in the gaseous state) with twice the volume of another gas, is said to have valence 2, etc. In practice, valences higher than 7 have not been found.

It becomes more complicated, when the combining substances *both* have valences higher than 1. But, if the valences are equal, then, we will usually find our simplest equation (univalent case):

$$A + B \rightarrow AB \quad \text{(Note: it is not possible to know the valence of A or B from this 1:1 equation.)}$$

When the valences differ, it becomes more interesting; for example, if A has valence 3 and B of 2. Thus, A has the ability to combine three times with another substance, and B twice its volume. Now, the equation is:

$$2A + 3B \rightarrow A_2B_3$$

In all these cases the astonishing fact is that the volume of the resulting compound is *smaller* than the total of the reacting gases. Experience has confirmed this.

We can make this clearer if we keep in mind that the reduction of volume is an essential characteristic for the combining of two gases:

$$A + B \rightarrow C$$

2 volumes $\rightarrow$ 1 volume

. . .

$$A + 7B \rightarrow AB7$$

8 volumes $\rightarrow$ 1 volume

The greater the valence of a substance, the more it can "be penetrated" or the more volume of another gas it can "take up" in its volume. The ability to condense volumes increases as valence increases.

$$2A + 3B \rightarrow A_2B_3$$

5 volumes $\rightarrow$ 1 volume

In these cases the proportion of initial volume to final volume, is equivalent to the sum of the valences of reactants to one. What happens when the valences are not mutually divisible (as in the last example)? What if A has valence 4 and B valence 2. Then, we may expect the following:

$$2A + 4B \rightarrow A_2B_4$$

Although there are examples of such cases, the following is found just as often:

$$A + 2B \rightarrow AB_2$$

Now, we can set up the following series of reaction equations:

valence A		valence B
1	$2A + B \rightarrow A_2B$	2
2	$2A + 2B \rightarrow A_2B_2 \quad A + B \rightarrow AB$	2
3	$2A + 3B \rightarrow A_2B_3$	2
4	$2A + 4B \rightarrow A_2B_4 \quad A + 2B \rightarrow AB_2$	2
5	$2A + 5B \rightarrow A_2B_5$	2
6	$2A + 6B \rightarrow A_2B_6 \quad A + 3B \rightarrow AB_3$	2
7	$2A + 7B \rightarrow A_2B_7$	2

A particularly useful aspect of this method is that we can let the students write out the rest of the series that comes about using other combinations of valences (say, B=3). Moreover, we have had ample opportunity to have them assist us—at least apparently—in working out and writing down the series. If we write out the series of equations properly, then they will reveal a clear and even rhythmical structure. We can then easily use them as material for homework exercises; the students can "play" with them as it were.

Furthermore, this method has the advantage that so far everything has been purely rational and, therefore, transparent. There are no unexpected exceptions which might arise from the unique properties of one or another substances, which could only be known from laboratory experience.

One of the reasons why the above discussion, which really does present the basis for the use of formulas, often is passed over, is that the reactions between gases are usually more complicated. It is certainly difficult to find examples which conform to the given mode of reacting [with elemental substances]. For example, we can bring iodine and mercury together in an intensively heated space, and the following reaction will then occur:

$$Hg + I \rightarrow HgI$$

An example known for a different case is the reaction of hydrogen and chlorine gases. One volume of hydrogen gas will react with an equal volume of chlorine gas to produce two volumes of hydrochloric acid gas. The reaction follows the scheme:

$$A + B \rightarrow 2C$$

This mode of reaction seems to stand in apparent contrast to the one developed by us. However, there is another possibility. We do not have to abandon our notion, if we can assume that in *this* case, A and B are *not* simple substances, but have already experienced a condensation previously. Then, A should be written as A_2 and B as B_2.

$$2A \rightarrow A_2$$

$$2B \rightarrow B_2$$

During the combining both substances could first expand to their double-volume and only then combine with the other. The reaction would only apparently proceed as follows:

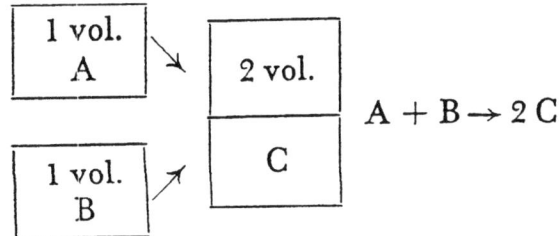

In reality, it occurs like this:

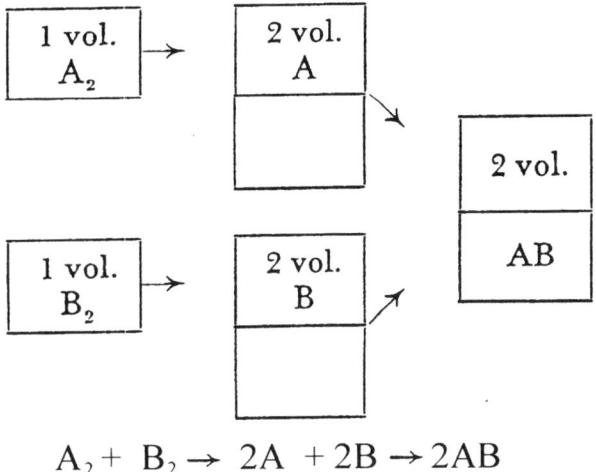

$$A_2 + B_2 \rightarrow 2A + 2B \rightarrow 2AB$$

This idea has been verified in practice and seems to conform with the facts and thus explains them. So we don't need to disregard our initial train of thought.

The reaction between hydrogen gas and chlorine gas proceeds according to this scheme

$$A_2 + B_2 \rightarrow 2\,AB$$

$$H_2 + Cl_2 \rightarrow 2\,HCl$$

Not only have we saved the line of reasoning we have been developing so far, but by explaining even this case, we have surprisingly given it more support.

In the above example, we have a reaction between two substances with valence 1. In the case of the formation of water from hydrogen gas and oxygen gas, we are dealing with substances having valences 1 and 2 respectively. We get the following scheme:

$$2\,A_2 + B_2 \rightarrow 4\,A^* + 2\,B^* \rightarrow 2\,A_2B$$

$$2\,H_2 + O_2 \rightarrow 2\,H_2O$$

When doing the experiment, we must take care that the water remains a gas by allowing the reaction to proceed at a temperature well above 100°C.

All this doesn't only apply to chemistry teaching, but I dare to maintain that these are actually the sequence of concepts that underlie the system of chemical formulas. These have always been used without quite realizing this. People thought that atoms and molecules were necessary to describe the phenomena of reacting volumes precisely and to explain them. In reality, there has been no discovery during the whole of the 19th century that drew attention to atoms or that necessitated atoms for their description. In many cases, we are dealing with gas volumes, while thinking we are using the concept of "the atom." The formulas of liquids and solids are based on the same concepts.

If the occasion arises, it is possible to extend this schematic preparation in the use of formulas still further. For example, an overview of the formulas of the most important acids, bases and salts can be given. It is quite all right to do this, as the students already know the most important examples.

When sodium is exposed to moist air, the result is:

$$4\,Na + O_2 \rightarrow 2\,Na_2O \quad \text{sodium oxide}$$

$$Na_2O + H_2O \rightarrow 2\,NaOH\ (Na_2O \cdot H_2O) \quad \text{sodium hydroxide}$$

If calcium is left lying in moist air:

$$2Ca + O_2 \rightarrow 2CaO \quad \text{calcium oxide}$$

$$CaO + H_2O \rightarrow Ca(OH)_2 \; (CaO \cdot H_2O) \quad \text{calcium hydroxide}$$

In general, the metals only produce a base if their valences are 1, 2 or 3. Thus, the most important hydroxides may be written as:

Me valence 1: MeOH $(Me_2O \cdot H_2O)$
Me valence 2: $Me(OH)_2$ $(MeO \cdot H_2O)$
Me valence 3: $Me(OH)_3$ $(Me_2O_3 \cdot H_2O)$

In general, substances produce acids when their valence lies between 3 and 7. If we express an acid-forming substance by the symbol Af, we get the following:

Af valence 3 (trivalent): $HAf + 3O_2 \rightarrow 2 Af_2O_3$

$$Af_2O_3 + H2O \rightarrow 2 HAf O_2 \; (H_2O \cdot Af_2O_3)$$

Af valence 4: $Af + O_2 \rightarrow Af O_2$

$$AfO2 + H_2O \rightarrow H_2Af O_3 \; (H_2O \cdot AfO_2)$$

Af valence 5: $HAfO_3$ $(H_2O \cdot Af_2O_5)$
Af valence 6: H_2AfO_4 $(H_2O \cdot AfO_3)$
Af valence 7: $HAfO_4$ $(H_2O \cdot Af_2O_7)$

Formulas for metal-salts of acids:

Me valence 1: Me Af O_2 Af valence 3
               Me_2 Af O_3 Af valence 4
               Me Af O_3 Af valence 5
                          etc.
Me valence 2: Me $(Af O_2)_2$ Af valence 3
               Me Af O_3 Af valence 4
               Me $(Af O_3)_2$ Af valence 5
                          etc.

Endnotes

[1] *Faculty Meetings with Rudolf Steiner*, GA300a, Apr 24, 1923, p605.

[2] Steiner, R. *World of the Senses and the World of the Spirit*, Lect. 4, p64; GA 134.

[3] *Goethe's Scientific Work*, Vol IV-2, notes by R. Steiner.

[4] Lommel, E. von, *Lehrbuch der Experimentalphysik*, Leipzig 1913 (and later), Introduction.

[5] Steiner, R. *Study of Man*, Lecture 3; GA 293.

V

Weight Ratios in Chemistry

The Balance and Materialism

We often say that modern chemistry began with Antoine L. Lavoisier (1743 - 1794). Lavoisier made much more exact use of the balance than his predecessors. By using the balance in this way, he discovered the role of oxygen during burning and rusting, for example, at least as far as this can be determined quantitatively. With this he created a foundation for the very precise chemical concepts we use today. But it is not realized widely enough that once and for all he brought an end to an epoch that had its own character and merit in relation to chemistry. Nobody will be sorry that he abolished the phlogiston theory, but it is of concern that his work increasingly fostered a one-sided view-point— everything now starts to be judged from the point of view of weight, of gravity. And, we have forgotten "levity" as the opposing force. Because Lavoisier (with others) worked towards a new, clear nomenclature, he contributed greatly to a real clarification; but he also aided the elimination of the last vestiges of a very old mystery-wisdom, which had still lived on in alchemic terminology. Whoever knows something about these intricate historical traces knows that this deed was necessary, but it did mean the end of a more imaginative, spiritual way of contemplating nature. From then on, chemistry in particular provided a firm basis for materialism, whose influence was steadily increasing.

Lavoisier based his method on a statement which was called the "Law of Lavoisier," (although others had expressed this before him) and is now called the Law of Conservation of Mass which is:

The total mass of the reacting substances is the same before and after the chemical reaction.

As a working-method (which is always used with reservations), nothing can be said against this statement, because it forms a sound and exact foundation. However, such a statement becomes questionable when it is transformed into something much more far-reaching, as the materialists did. Ernst Haeckel (1834 - 1919), for example, the greatest champion of materialism, wrote in his book "Die Weltraetsel" (1899): "The sum of all matter, which fills infinite space, is unchangeable." He attributes this to Lavoisier, despite the fact that this is something different from what is expressed in Lavoisier's law. What started as a *hypothesis* grew to enormous importance, to a fundamental "law" for *all time* and *the whole universe*. The materialists declared matter to be an eternal principle and thereby renounced the idea of a continuous process of divine creation.

The law of Proust (1754 - 1826), or the Law of Constant Proportions, is seen as a second foundation of chemistry. This law may be stated as follows:

Substances combine with each other only in definite mass ratios.

This and other events all connected with the development of chemistry, and all taking place around 1800, can be seen as symptoms that a philosophy already prevalent in astronomy, mechanics, mathematics, and physics in the previous centuries, now was also starting to completely dominate chemistry as well.

Scientists who embraced this philosophy found their inner certainty through mathematical concepts and tried to understand and even dominate nature with their help. Through them, the approach of evaluating processes in nature according to measure, number, and weight gained an ever-increasing importance; we might even speak of steadily increasing overestimation of the quantifiable. On the contrary, they showed little or no interest in the qualitative aspects. They didn't want to recognize the qualitative as something that could have scientific meaning. On the contrary, they viewed (and nowadays still view) this side of nature as something subjective, that at most can serve as a signal, and also felt that we should always look for something "objective" (quantifiable) behind it. Sound is in essence is considered "nothing more than" vibrations of air; light, "merely" a stream of minute particles or waves in a hypothetical ether. In every realm, this fostered an inclination to reduce the qualities that we observe to mechanical causes.

The atomic theory is an attempt to eliminate all the phenomena of chemical reactions as unimportant for the senses, and to replace them by a mechanism of atoms. Dalton, for example, was confronted with the observations that two substances completely interpenetrate each other when they combine chemically, and even, to a greater or lesser extent, when two gases mix (as in the atmosphere), or when a gas dissolves in a fluid. Among other things, his theory is an attempt to answer the question: How can we understand the phenomenon that two substances interpenetrate, when we assume that they themselves are conceived to be impenetrable (solid matter)? He found the solution in concepts which were already current at that time, according to which every substance [is thought to] consist of minute particles, separated by fairly large distances. In the case of two mixing gases, the atoms were supposed to race past one another without influencing each other. In the case of gases dissolving in water, something similar was supposed to happen. When two substances combine chemically, this would mean numerous "connections" of pairs or groups of particles. The substances themselves, as atoms, weren't supposed to be changed, but only the ways in which the minute particles could move. Of course, this was a sort of explanation of the Law of Lavoisier. According to Dalton, during a chemical reaction the "particles" remain unchanged, only the way they combine changes.

Dalton could now explain the fixed weight ratio of a compound directly from his theory of combined particles. Conversely, from the weight ratio of the compound we could learn something about the combining ratio of the particles. Of course, these were too small to be weighed in grams, but we could express their weight in a certain number, if in accordance with the indicated method, we compared them with the lightest particles known (with the hydrogen atom). Dalton was the first who made a list of "atomic weights" (nowadays we would call them "equivalent weights").

The Law of Multiple Proportions, which is attributed to him, too, can also be easily explained by his theory. If a substance A can combine in different proportions with a substance B, then we can ask ourselves, how the different amounts of substance B, which combine with the same amount of substance A, relate to each other. Theoretically, the relationship could be any ratio; but experimentally, this is not the case. The amounts of substances always relate to each other as small whole numbers.

From this Dalton immediately developed the idea that the amounts behaved more or less as sharply defined entities. If a chemical compound by nature consists of a bound sequence of two or more particles, then such simple relations must follow. Dalton is often spoken of as a genius. Against this we can say that he must have been a great illusionist above all. The world of phenomena is so rich and magnificent and full of remarkable relationships, that we must be somewhat conceited if we believe we can reduce all this to a few simple mechanical laws. This approach was perhaps easier for him, because an important part of the abundance of the phenomena was lost to him: he was color-blind.

Somewhat later, when J.J. Berzelius (1779 - 1848) finished his monumental work in which he examined the composition of the most common compounds and most importantly determined the combining weights of all known elements, it seems as if he is in agreement with Dalton. He even voices his admiration for the atomic theory. In actual fact, he goes about his work in a much more unbiased way than Dalton and so leaves more possibilities for another point of view. He knows very well and even mentions that he doesn't need atoms for his work, but in actual fact is dealing with gas volumes. He even talks directly of a "volume theory" and has the opinion that this is better supported by the facts than is the atomic theory. To reconcile both theories he voices the opinion that under the same conditions, equal volumes of *different* gases contain the *same* number of atoms. This is nothing other than the law Avogadro stated later. Proposing this law (which for the moment should only be regarded as an hypothesis), it follows only that we thought we were thinking in terms of particles, while in reality we are dealing with volumes. In these reflections and calculations, an 'atom' is mostly nothing other than a definite volume of a gas. (We use the thought "atom" to "explain" the observation of whole number ratios.)

Berzelius was also the inventor of the formula script as used today, and in so doing has contributed enormously to the order, precision and clarity in chemistry. He stresses that every letter in his formulas really represents an equal volume of gas, but, of course, with a different weight. These weights stand to each other in the same ratio as the "atomic weights." It is interesting that in determining the respective atomic-combining weights, he based everything on oxygen, since most compounds are oxides.

It is amazing how long it took before the atomic weight, in Berzelius' sense, really was used as a common basis for quantitative calculations, and how much effort it cost to clearly distinguish the concepts "atomic weight," "molecular weight," and "equivalent weight" from each other.

The Balance and the Spiritual Ordering of Nature

When we review the history of chemistry in this way, it becomes clear that the powers which want to steer humanity in the direction of materialism stood actively in the background during the exploration of weight ratios in the 18th and 19th century. To protect ourselves against these dangers, it is best to turn to a path of reflection that is directly aimed towards the spiritual.

According to old traditions, shortly after the autumn equinox the sun moves into the constellation Libra (Latin "the scales"). Immediately after entering the dark season [late October], the Scales are put before us as an admonition. Not long after the beginning of autumn and starting around the time the sun is in Libra is the season of "Michaelmas," in which we think of St. Michael the great dragon-fighter, often depicted holding a balance.

What does it mean that Michael, the leader of the arch-angels, who always wants to keep the view into the spiritual world clear, free from confusion and shadows, not only conquers the dragon but also holds a balance-scale in his hand? Could it be possible to find a quite different, more spiritual view of the balance, which would leave open a path to the spirit in the field of chemical reactions?

The first question we must address is: when the weight of the substances remains the same before and after a chemical reaction, is this proof that we are always dealing with the same substances, even if in a different form? The answer must be no. We may just as well assume that substance disappears and new substance appears, but in such a way, that the weight remains equal.

If we reason along the lines of the conservation of mass, then the whole process takes place, as it were, on one pan of the balance, while on the other the weight lies patiently. The balance then remains undisturbed, and we are led to the belief that nothing radical has occurred.

However, if substance actually disappeared and new substance was created, then there has to be an "ordering principle" in nature for which the moving balance is an ideal image. If one pan of the balance sinks, the other will rise in the same proportion. A definite amount of substance cannot disappear without the same weight of other substances appearing.

We must now briefly address a subject that is always brought up in conjunction with the above-mentioned point, and where we have a similar problem. That subject is the "conservation of energy."

It was, of course, known for a long time that we can generate movement through heat and heat through movement. Only gradually did it became clear that we can convert all forces of nature into one another, and in fact in certain proportions that are measurable and can be expressed in numbers. The credit for discovering such relations between the forces of nature belongs to J. R. Mayer (1814 - 1878). He expressed the equivalence that exists between heat and movement for the first time numerically, using specific units.

To be able to put these and similar discoveries into thoughts, the concept of "energy" was developed. Now it was possible to express the value of the effect of every force of nature as a unit of energy.

Energy and equivalence belong together. Energy is the measure in which we express the equivalence. The forces of nature can express themselves in different ways in the world of experience. But, as soon as we talk about their "energy," then we are speaking about that in which they are equivalent, disregarding their individual qualities.

Rudolf Steiner says in a seldom-mentioned passage: "The phenomena of the observable world/nature, light, heat or warmth, electricity, magnetism, etc, can be brought under the general concept of mechanical work done (effective force), that is, energy. When light, heat, etc. bring about a change in a body, then this is (construed as) work done (by some force). By describing light, heat, etc., as "energy," we have disregarded all the details which are specific for the various senses, and only taken into account a general property, common to them all."[1]

In a way, in terms of "energy," it makes no difference if it occurs as electricity, heat, or light. Energy is, therefore, a common unit for forces, in the same way that money is for the exchange of goods. Here also—as often is the case—we see the tendency to think the opposite to truth. We imagine that energy is the real thing, and the acting forces of nature only its manifestations. In that case we can really speak of the conservation of energy. However, we should understand that there is a sort of interplay going on between the forces of nature. We can only create a certain amount of heat, if there is a certain amount of chemical energy or mechanical movement or electricity available and we want to use it. It is not a question of conservation of energy, but one of *exchange* of forces of nature. The one is created, while the other vanishes.

(Note that today, achievement, *work* (done), output, efficiency, (electrical) power, wattage, (mech.) brake horse-power, all these concepts are distinct from: *energy*; work/time = Newton-meter/second = Watt, or Joule/s, energy = work, a certain effort in a certain time, not per unit of time; this last is power = J/s. D.R.)

We have reached a realm where heat takes on a sort of key role. Therefore, it should not surprise us that Rudolf Steiner tried very hard to develop a deeper insight into the behavior of heat. He strongly emphasized that it is not only a case of emerging into "observable/phenomenal" nature and disappearing again, but also of transitions from spatial to non-spatial states and vice versa. It is necessary for the further development of science that the general interaction of the spatial and non-spatial be given more attention.

We now return again to the weight ratios. As we saw previously, it is one of the most remarkable phenomena that during a chemical reaction the "impenetrability" of matter is overcome, and two different substances become a complete new unity. Thus, we have found occurrences in the realm of the material, which follow a fundamental law of the spiritual.

Other, equally radical phenomena are connected with this. The properties by which we recognize the substances taking part in the reaction disappear, and totally new substances with new properties emerge. Only one property, the weight (mass), has remained the same.

However, is this sufficient grounds to believe in the conservation of matter in the sense that we assume that we are dealing with the same substance before and after the reaction? Let us see what we can find in a student textbook, wherein the authors were very particular in their wording.[2] First, they note that the Law of Lavoisier had been stated earlier; then they continue as follows: "This law is also known as the law of the indestructibility of matter and considers under 'substance,' everything that has weight." Therefore, this word has quite a different meaning than normally associated with it in chemistry. Substances are recognized by their properties, but these disappear during a chemical reaction; which means that the "chemicals" themselves vanish and new ones appear in their place. The total weight (more precisely: the mass) of the substances, thus "the substantiality," remains, however. So "Law of the conservation of *mass*" would be a more accurate description of the aforesaid law.

From this really very precise wording, it becomes very clear what the real issue is. If a specific amount of sulphur is used to make iron sulfide and then as much sulphur as possible is recovered from the iron sulfide through various reactions, then the initial weight of the sulphur will be equal to that of the sulphur recovered. However, it is impossible to maintain that it is the same sulphur (from a phenomenological philosophical viewpoint).

We stand here at a similar junction as we encountered with the definition of "impenetrability." If we emphasize "conservation" when discussing a compound, then we are headed for materialism. However, if we focus on "becoming and disappearing," then we keep all the possibilities open for a mode of reflection which takes the spirit into account, or can even recognize that spirit is primary to matter.

In any event, it is desirable to be clear that we stand here before a threshold. There is something which initially presents itself to our perception, but that later withdraws again. Simultaneously, something different comes into our field of vision, which wasn't there initially.

The Spiritual Background of Materialism

We can only really understand what has been discussed so far, in a more characterizing and descriptive manner, if we have experienced through Spiritual Science as given by Rudolf Steiner, that a mighty battle lies at the foundation of the cosmic order. Certain powerful spiritual beings have disengaged themselves from the divine harmony and have rebelled against the proper divine powers. This battle has ended, above all where natural creation is concerned, in the victory of the true divine powers. By way of comparison we could speak of a truce being signed after the victory. A large part of the contents of that peace treaty we can gather from the existing order of nature. In human consciousness, the battle continues. Indications of this are to be found in many traditional tales. It is always Michael under whose leadership the battle is fought, and in the end the rebels are rejected from heaven and continue their activity as "dragons" on earth. The conquered powers, who are also known by the name of Satan, have been exiled from the spiritual to the spatial world. There they have their task. As far as nature is concerned, they must subject themselves to the order as it was once laid down. But there is no treaty that stops them from getting the human consciousness under their influence; they have a free hand here.

The divine powers have completely retreated from the realm of space. Therefore, they can neither interfere or exert an influence by using forces of nature. They live in and rule a realm that is supersensible in nature. It is clear that with the establishment of the order of nature, the period of creation of the world ended.

A remarkably well ordered equilibrium establishes itself at the boundary between these realms. In the very beginning of creation, spirit continually condensed to matter, and thereby was estranged from itself. In that way, in the midst of the non-spatial supersensible world, the spatial world slowly evolved. However, the world remained one—one part cannot exist without the other. Even now, non-spatial becomes spatial. This, however, is directly compensated by spatial becoming non-spatial. If this weren't the case, Satan's kingdom would become ever larger. This equilibrium is part of the above mentioned peace treaty. We could just as easily speak of a contract. And, as in the case of a proper contract, there are precisely determined ratios of values, which can be expressed in numbers. In the case of the forces of nature we can come to an image of these ratios of values, by using the concept of energy. If a force of nature, which represents a certain amount of energy, vanishes, then other forces of nature must come into play, which together represent the same amount of energy. With regards to the chemical reactions, the vanishing and emerging substances must have the same mass.

If we view the order of nature as the result of a battle, then we can also understand how the battle continues that is fought in human consciousness. It is very important for humans and for the universe that they form a true picture of the relation between spirit and matter. However, they also do have the freedom to embrace any errors or misconceptions. And Satan uses this freedom. With all his might he tries to influence people in such a way that they restrict their observations and also their thoughts to the realm that is under his domination. He wants to create this into a kingdom for himself, in which as many people as possible are enslaved. Instead of accepting that a contract is a matter of give and take, he exclusively represents his own rights. He boasts about the fact that he has so much energy, so much mass at his disposal, that cannot be taken away.

If we think in terms of conservation of matter and energy, we are being influenced by him. However, if we want to evade his influence, then we must fight a rigorous personal battle, because to start on the path of truth, we have to rid ourselves of very pervasive and, in a certain sense, very seductive forms of thought.

At the start of our last discussions we placed the battle of Michael with the dragon. Michael is not only the great leader in the battle, but also the guardian who scrupulously guards the threshold between the spiritual and material domains. The scales in his hand point to the fact that the harmony he guards rests in every respect on equilibrium. However, the dragon doesn't speak of equilibrium but possession, a right he wields. Therefore, he can deceive himself and others that he is firmly entrenched in his possessions and that he will lose nothing, because he hasn't any interest in the differences between substances and forces of nature. That something changes *qualitatively* during a chemical reaction doesn't bother him in the least as long as the *quantity* remains the same. That he has obligations and that he even fulfils them, he tries to hide; he focuses all attention on an illusory law.

At any rate, the scales in Michael's hand indicate that the exact harmony that can be found in the area of equilibrium is also a result of his victory. Everything that belongs to this area must be treated in such a way that the harmonious relations and, therefore, the great all-encompassing structure immediately comes into view. Only when we treat it in this way will the perspective of the spiritual origin of nature remain free.

Equivalent Number as a Pure Expression of Weight Ratios

In directing our attention to the manipulation of weight ratios in chemical reaction, we must first of all start from experience. The historical train of thought that formed the starting point for setting up the writing of chemical formulas must remain in the background.

As an example, we choose carbonic acid gas (carbon dioxide), a compound of carbon and oxygen, CO_2. We find the ratio:

$$\text{weight C} : \text{weight O} = 3 : 8$$

At the moment, we are not concerned with the specific numbers given above, but with their *ratio*. We could just as well have taken the numbers 12 and 32, or 1.5 and 4.

As the research methods became more precise, it became increasingly clear that such ratios are exact and invariant. Initially people doubted this; but when it was proved repeatedly to be the case, the opinion grew that it had to be established as a law. This is "Proust's Law," or the "Law of Constant Proportions":

Every compound has a constant composition, or *Substances will only combine in definite weight ratios.*

This law has indeed two aspects. Initially it seemed the focus was primarily on the question to what degree small deviations can occur in the composition of compounds. Later, it became more and more clear from research that between two (and also between more than two) substances that react with each other, there exists a specific weight ratio for those substances. But it never concerns only a few substances; with each ratio every other ratio is taken into account. There exists, therefore, *a system of ratios* that encompasses all substances. We could speak of a *network of weight ratios.* When a pair of substances react with each other, a fragment of this network is made visible. Of course, everybody knows this very well, but most forget to express this clearly in words. The only time it is mentioned is when we are introducing the concept of equivalent weight. How do we get a clear picture of this network? We take one of the substances as starting point and relate all other substances to it. We get a definite number for every substance. This number plays a fundamental role in all compounds of a substance, not, of course, on its own but in relation to the numbers of other substances. It is obvious to choose oxygen as reference substance, as Berzelius did, because it plays a central role in chemistry. However, hydrogen was chosen as unity, because it has the smallest number. All other numbers are, therefore, larger than one.

However, we must be aware that we determine a regularity, which isn't concerned with specific numbers, but with specific *ratios* of random numbers, although expressed in the form of numbers. To describe this network, we resort to an assigning which does not correspond 100% with reality. In setting up the standard we take a further step in choosing a specific amount of hydrogen, namely 1 gram as reference. For all other substances the weights will be in grams, which are called equivalent weights. The equivalent weight of an element is the amount of substance expressed in grams, that will react with 1 g of hydrogen (through combination, substitution, or displacement).

As will be clear from the foregoing, in introducing the concept of equivalent weight, we have left out the concept equivalent ratio, as such. This gives rise to a purer description of nature, because nature does not calculate in grams. I'm of the opinion that the concept "equivalent weight" could well be called "equivalent number." This number expresses the weight ratio between a quantity of a substance and that of hydrogen (before 1961; afterwards, based on 12.0 g of C^{12}) with which the substance will react with or can displace.

We will give a few examples, which will clarify this further.

We first express the weight ratios in arbitrary numbers:

$$CO_2 - \text{weight C : weight H} = 1.5 : 4$$
$$H_2O - \text{weight H : weight O} = 8 : 64$$
$$CH_4 - \text{weight C : weight H} = 7 : 21$$

Now we express these ratios with the help of "equivalent numbers."[3]

$$H_2O \qquad - \text{weight H : weight O} = 1 : 8$$
$$H_4C\ (CH_4) \quad - \text{weight H : weight C} = 1 : 3$$
$$CO_2 \qquad - \text{weight C : weight O} = 3 : 8$$

In this way part of the network of ratios becomes visible, because every substance has its "standard number" (equivalent number).

If it weren't for complications, which could bring confusion into the simple relationships we have developed so far, we could use very simple chemical formulas. Water would become HO, carbon dioxide CO, etc., whereby every symbol would be representative of its equivalent number.

The complication involves substances with more than one equivalent number, which is a consequence of the fact that various substances can combine with different quantities of another substance, though still in definite whole-number ratios. A good example is sulphur:

$$\text{S - valence 2, e.g. in } H_2S - \text{equivalent number: } 16$$
$$\text{S - valence 4, e.g. in } SO_2 - \text{equivalent number: } 8$$
$$\text{S - valence 6, e.g. in } SO_3 - \text{equivalent number: } 5.33$$

This means nothing less than that it isn't possible, without encountering logical gaps, to introduce chemical formulas, by way of weight ratios. To start as we did, from volume ratios, is the logical way.

So now, we have to find the transition from volume ratios, to weight ratios.

Gas Volumes and Weight Ratios

The most natural transition from gas-volume ratios to weight ratios is found in the phenomena which we find designated by the word "density." Every substance, every solid, gas or liquid, has a specific density. We could call it the degree of submission to gravity. Moreover, density is mainly a relative characteristic.

With solids and liquids, to indicate density we can use the concept "specific gravity." We determine the weight of 1 cm³ of water at 4°C and use that as unit weight = 1.00. Then, we determine the weight of 1 cm³ of the substance to be examined. If this turns out to be 6 g, then the density of this substance is 6 times that of water (specific gravity = 6). In the case of gases and vapors it isn't possible to express the density in the form of the relative density. In the first place the resulting numbers would be very small in other words the unit is to large.[4] Secondly, a gas behaves quite differently with regards to its density than do solids or liquids, where the density is hardly dependent on the pressure and only slightly on the temperature. With gases the density varies greatly with changes in temperature and pressure. The degree of change, however, is nearly the same for all gases. We can use this fact by taking hydrogen as reference gas, because it is has the smallest density, and comparing the weights of the other gases under similar conditions of temperature and pressure to hydrogen. This results in a number for every gas. The number is called its "vapor density."

$$D = \frac{\text{weight of a volume of gas}}{\text{weight of the same volume of hydrogen}} \quad \text{(same p,T)}$$

(temperature and pressure = T,p)

The vapor density may be treated in the same way as the equivalent weight. As seen from the reality of nature, the numbers are only significant when seen in relation to other numbers. Just as there is a ratio of the densities of oxygen and nitrogen, so there is one between oxygen and hydrogen. The first ratio is expressed by the numbers 16 and 14, where each number expresses the ratio in relation to hydrogen. However, it could have just as well been expressed by the numbers 4 and 3.5 etc., by disregarding the ratio to hydrogen.

If a chemical reaction takes place between two gases which combine in equal volumes, then the ratio of the densities expresses the ratio of the weights of the two substances.

Taking the reaction at an elevated temperature between mercury and iodine vapor (already discussed in Chapter V), we find the following:

1 volume Hg(g) + 1 volume I(g) → 1 volume HgI(g)
density: D = 100.3   D = 63.5   D = 100.3 + 63.5
weight Hg : weight I : weight HgI = 100.3:63.5:(100.3 + 63.5)

The density of a compound is given by a number which equals the sum of the numbers, which designate the densities of the participating substances. That is exactly what we would logically expect, because combining in fact implies condensation:

$$A + B \rightarrow AB$$
weight A : weight B : weight AB = x:y:(x+y)

In most cases, however, it works differently. As an example, we'll again take the reaction between hydrogen and chlorine:

$$H_2 + Cl_2 \rightarrow 2\,HCl$$
$$D = 1 \quad D = 35.5 \quad D = 18.25$$

weight H_2 : weight Cl_2 : weight 2 HCl = 1 : 35.5 : 2{(1/2)(1 + 35.5)}

We have already previously encountered the problem of how to explain this remarkable phenomenon. We saw what was first suspected and then later on verified, that hydrogen and chlorine normally do not occur in their simplest forms. We must see hydrogen as a compound of the substance H or as a condensation of H with itself; the same applies to chlorine with regard to the substance Cl.

$$2\,H \rightarrow H_2$$
$$2\,Cl \rightarrow Cl_2$$

These substances (H and Cl) are half as dense as hydrogen and chlorine. The vapor density of (elemental) H would, therefore, be 1/2 of the density of naturally-occurring hydrogen gas.

This means nothing less than that (natural) hydrogen is not the lightest substance and, therefore, should not be used as reference substance in determining densities. The ideal substance would be the [elemental] substance H. We must, therefore, correct, or rather multiply by 2, all vapor densities. In this somewhat unusual way, we arrive at the concept, which commonly is called the molecular weight mass.

$$M = \frac{\text{weight of a volume of gaseous substance}}{\text{weight of the same volume of substance H}}$$

From this point of view, the name "molecular weight (mass)" is very unsatisfactory. We could better call this concept absolute vapor density. Perhaps it is even necessary to give the one concept two names.

It is self-evident that for those elements in gas or vapor form (which are their least-condensed state), we have determined the vapor density and have calculated their molecular weights. We then obtain a weight-ratio of a gaseous substance to hydrogen, both in their least dense states. People even found it worthwhile to determine this weight-ratio for all known elements, although in many cases it wasn't possible to do this directly, because not all substances could be put into a gaseous state, or for other reasons.[5] The resulting numbers are called "atomic weights" (or "elemental weights"). We could thus characterize the atomic weight as the absolute vapor density of a substance in its least dense state.

Up to now we have only discussed a certain aspect of the atomic weight, namely gravity in the gaseous state. However, the atomic weight also comes into play as a ratio (of numbers) when substances react. To that degree, the atomic weight is related to the equivalent number. We can easily find the atomic weight of an element by multiplying the equivalent number by the valence. For substances that have valence 1, the equivalent number and the atomic weight are (numerically) the same. For substances that have valence 2, the equivalent number is half of the value of the atomic weight. Substances that have more than one valence have different equivalent numbers but only one atomic weight.

The atomic weights point in the same way to a net of numbers as do the equivalent numbers. Here also the numbers themselves are not so very important, but the ratios between them are. As every element has only one atomic weight, the net of these is simpler than that of equivalent numbers.

When we are writing formulas, which are based on weight ratios, we can either use equivalent numbers or atomic weights. Using equivalent numbers would make the formulas very simple. However, we get into difficulties—as we have seen—by the fact that sometimes an element can have various equivalent numbers. However, if we use atomic weights this immediately implies that we are using the volume ratios, which does lead to greater clarity. We must, however, accept that the atomic weights appear as multiples in the weight ratios.

As an example, we'll take the equation for the burning of sulphur to sulphur dioxide. First of all, we'll set out the equivalent numbers and then the weight ratios based on the atomic weights:

$$\text{sulphur} + \text{oxygen} \longrightarrow \text{sulphur dioxide}$$
$$\text{equivalent numbers} \quad 8 : 8 : (8+8)$$
$$\text{weight ratios based on atomic weights}$$
$$32 : 2\times16 : (32 + 2\times16)$$

The last line is more complicated, but, as we know from experience, in the end clearer than the first.

We want once more to return to the molecular and atomic weight, but now following the usual approach.

We start with the vapor density:

$$D = \frac{\text{weight volume of gas}}{\text{weight same volume hydrogen gas}} = \frac{\text{weight } x \text{ molecules of gas}}{\text{weight } x \text{ molecules hydrogen}}$$

$$= \frac{\text{weight of 1 molecule of gas}}{\text{weight of 1 molecule of hydrogen}}$$

The determination of the vapor density could be used without further calculation to give the weight ratio between a molecule of a certain gas and that of a hydrogen molecule.

When we are already looking for a unit to express the discovered quantities in, then it is better to take the smallest particle than the next largest. This smallest particle is a hydrogen atom. Thus, we arrive at a weight ratio, which is called the molecular weight:

$$M = \frac{\text{weight of 1 molecule of a gas}}{\text{weight of 1 atom of hydrogen}}$$

Without doubt we had the feeling that setting up the reasoning in this way, by starting with a sharply defined particle, would prove a firm foundation.

However, it wasn't always clear that, in reality, the vapor density is the best guiding principle and that by introducing the theoretical concept of an "atom" or "molecule," (phenomenologically-true) number ratios were being exchanged, for (theoretical) material mental concepts, in order to make them more concrete.

Thus, nobody can dispute our right to reverse the argument and start with the weight of a certain volume of hydrogen gas in the atomic state, notwithstanding the fact that nobody would be capable of producing this substance in such a state, that a certain volume could be weighed. When we say, as has been previously mentioned:

$$M = \frac{\text{weight of a certain volume of a substance in a gaseous state}}{\text{weight of the same volume of H}}$$

As always, under the same conditions of temperature and pressure then, our unit may have a more or less hypothetical character, but at any rate to a lesser extent than the notions of atoms and molecules did at the time they were introduced. We can be sure that this procedure would conform to reality, if we would be able to produce the substance H as a real gas.[6]

For completeness we would like to show the following:

$$M = \frac{\text{weight of a molecule of a substance}}{\text{weight of 1 atom of hydrogen}} = \frac{\text{weight of } x \text{ molecules}}{\text{weight of } x \text{ atoms}}$$

$$= \frac{\text{weight of a volume of substance as a gas}}{\text{weight of a volume of hydrogen gas in state H}}$$

Thus, along this path we arrive at taking molecular weight as a sort of "vapor density." We could also say: the molecular weight refers to a weight ratio which emerges as soon as the substance is brought into the gaseous state.

For atomic weight, the same applies, as this is not essentially a different case from molecular weight.

$$A = \frac{\text{weight of 1 atom of an element}}{\text{weight of 1 atom of hydrogen}} = \frac{\text{weight of } x \text{ atoms}}{\text{weight of } x \text{ atoms of H}}$$

$$= \frac{\text{weight of a volume of a certain element in gaseous form}}{\text{weight of the same volume of hydrogen in state H}}$$

When we bring a chemical element in its most simple form (thus, A_1 and not A_2 or A_3) into the gaseous state, and then compare the weight of one volume with that of an equal weight of hydrogen in the state H, the result is a ratio, which is called the atomic weight of the element.

It was often very difficult to state with any certainty: this is the atomic weight of an element—because we never knew whether we were dealing with a multiple thereof.

If, however, we have managed to determine the atomic weight, and we have the formula for a specific substance, then it is possible to find the magnitude of the molecular weight by multiplying the magnitudes of the atomic weights with their subscripts and then adding the results. We shall explain the procedure for the case of sulfuric acid. The formula is H_2SO_4. If it were in a position to bring this substance into the gaseous phase without it decomposing, and then to analyze it in such a manner that the products of analysis would remain gases at that temperature, this analysis would give the following result:

$$1 \text{ vol. } H_2SO_4 \rightarrow 2 \text{ vol. } H + 1 \text{ vol. } S + 4 \text{ vol. } O$$
$$\text{wt. 2 vol. } H : \text{wt. 1 vol. } S : \text{wt. 4 vol. } O : \text{wt. 1 vol. } H_2SO_4$$
$$= 2 \times 1 : 1 \times 32 : 4 \times 16 : (2 \times 1 + 32 + 4 \times 16)$$

All the various volumes of the components have been contracted to the one volume of the compound and contribute to the density.

If we were only aiming at a true rendering of the truth, then irrespective of the effort the application of these thoughts cost, we should try to remain in the realm of ratios. Every step in the direction of fixing these ratios makes us more rigid in the realm of concepts. If we are not conscious of this, then we are irrevocably led into materialistic ways of thinking from which it is very hard to extract ourselves.

With regard to a substance like carbon monoxide, for example, we should not develop any preference for the numbers 12 and 16, but be just as content with the ratios (3/4) or (9/12) as with (12/16).

In contrast, there is the discovery that the atomic weights stand in a special relation to hydrogen. If we only actually had the simple, completely pure chemical elements (hence, had separated the mixture of isotopes), then their atomic weights would, in fact, be whole numbers. We should only take the least dense form of hydrogen as a reference substance. The atomic weights of the elements are, therefore, multiples of that of hydrogen. The fixation of the relationships as numbers, where hydrogen is taken as a unity reference, is, therefore, legitimate to a certain degree.

However, if for oxygen also found in nature we fix the atomic weight at 16 and fix the reference weight for atomic weight calculation at one-sixteenth, then we introduce a hydrogen which has been "corrected" in the interest of practicality. The harmony of nature is pushed to the background for the benefit of our convenience.

When talking about gram-mole and gram-atom[7] we are completely moving away from pure science.

We read in a textbook: "A gram-atom of an element is the number of grams of that element which is equal to the magnitude of the atomic weight." This sentence isn't very well phrased, but it is much better than the one we found in an otherwise excellent university text: "A gram-molecule is the molecular weight expressed in grams." Of course, it is impossible to express the molecular weight in grams. In actual fact for the convenience of calculations we have, without any hesitation, placed grams after the magnitude of the molecular weight. We can possibly express this somewhat better as follows: It was found necessary to have fixed, comparable amounts of the different substances as a starting point for chemical calculations. The most obvious thing to do would have been to start off with the volume ratios of the substances as gases and then, for example, to use the weight of 1 Liter of every substance at a certain temperature. The weights of these quantities would, of course, have been in the ratio of the molecular weights. If the weight of 1 L H_2 (at 0°C and 760 mm) is 0.09 g, then the weight of 1 L of O_2 (under the same conditions) is 16x0.09 g, etc. This method is too difficult to use in calculations. We start, therefore, with a volume of 22.4 L (at 0°C and 760 mm). 22.4 L of hydrogen weighs 2 g; 22.4 L of oxygen weighs 32 g; 22.4L of iodine (I) at 0°C and 760 mm would weigh 127 g. In this way the weights correspond numerically to the atomic and molecular weights.

When a formula or a symbol is found anywhere in an equation, it usually stands for an amount of a compound or an element expressed in grams, which is numerically equivalent to the atomic or molecular weight. These amounts are called gram-atom or gram-molecule. The mole (=gram-atom and gram-molecule) is not a very clear concept. That they have been put forward is only justified because of the simplicity they offer for calculations. For example, a mole is a quantity that is based on two completely different notions. One is grounded in nature, the other is arbitrary.

Through much usage such concepts acquire a meaning for our thought life that is out of proportion to what they should have. We start to use them as natural facts and then use them to prove that all gram-atoms and gram-molecules contain the same number of atoms or molecules. From this we further deduce that for gaseous substances the volume of a mole always has the same magnitude under the same conditions. We thus prove what we stated initially.

We will take an example from a school textbook:

"If an element has atomic weight A, and the weight of 1 hydrogen atom is h gram, then:

$$A = \frac{\text{weight of 1 atom of element}}{\text{h gram}}$$

From this it follows that 1 atom of the element weighs A x h grams.

Because 1 gram-atom weighs A grams, it contains A grams/(A h grams) = 1/h atoms. This number is therefore independent of A."

In the same way we can argue that 1 gram molecule of a substance will always contain (1/h) molecules.

The argument then goes on as follows: "It was proved previously, that the gram-molecules of all substances contained the same number of molecules. In conjunction with Avogadro's law it then follows that for every gas the volume of 1 gram-molecule, at the same temperature and pressure, is the same for all gases".

If such thinking proves anything, it is only that very clear thinking can be exceptionally inconsequent, even muddled.

What is really the case?

atomic weight H : atomic weight O : atomic weight O_2 = 1:16:32

If 1 atom H weighs h grams, then 1 atom O weighs 16 x h gram and 1 molecule O_2 weighs 32 x h g. If I now replace h g, 16xh g and 32xh g by 1 g, 16 g and 32 g, then I have done nothing more than multiplied all three by the same number (1/h). I am not comparing the weights of an atom or a molecule, but at most of (1/h) atoms or molecules.

And we really don't need to have deduced again that a mole of every gas, at the same temperature and pressure, has the same volume. In comparing moles of different substances, we are comparing them according to their molecular weights. These have a direct relation with the vapor density; respectively, they are in a certain sense vapor densities and come about by comparing equal volumes.

If we want to develop concepts that are not only easy to use, but also remain as close to reality as possible, then it is above all important to make a clear distinction between science and technology. A large part of what we call science nowadays is really already technology. We do not first ask ourselves the question: Am I on the path to truth? But: How can I manipulate the facts as easily and as simply as possible?

In actual fact it is quite possible to dispense with the "mole" concept. However, because it is currently used, we will have to introduce it at some time. With the students we should first use pure ratios in calculations before starting with gram-molecule and gram-atom.

A simple chemical calculation nearly always proceeds along the following lines:

For example, we ask: How much oxygen is needed to burn 20 g of phosphorus? First, we set up the reaction equation, from which the ratios follow, thanks to previous research:

$$4 P + 5 O_2 \rightarrow 2 P_2O_5$$

weight P : weight O2 : weight P_2O_5 = 4x31 : 5x(2x16) : 2x(2x31 +5x16)

weight P : weight O_2 = 4x31 : 5x32

$$20 \text{ g} : x \text{ g} = 124 : 160$$

$$= \frac{20 \times 160}{124} = 25.8$$

However, if we write the following, then this leads on to a more materialistic consideration:

$$C + O_2 \rightarrow CO_2$$
$$12\,g + 32\,g \rightarrow (12 + 32)\,g$$

Atomism as a Starting Point for Explaining Natural Phenomena

So, we come to the question: Must this chapter be interpreted as arguing against the appearance of atoms in nature? I believe that I have not presented a single argument against the occurrence of smallest particles. What such arguments do show, however, is that the atomic theory as introduced by Dalton was superfluous and even in may ways misleading. There exists a great distinction between the atom as starting point for every explanation, and atomic particles observed as phenomena. The first is a product of our thinking, even if it is more or less made to correlate to certain phenomena. The second is a fact of nature.[8]

Precisely when we tend to think of the phenomena of nature in an atomistic way, as many people do nowadays, we should be thankful that the notions about their essence and position in the universe are based on clear thoughts and observations and not on all sorts of confusing constructions.

It follows from several chapters of this book that there exists a whole world of precisely ordered relations that can be discovered within the finer qualities of substances and that cannot be explained out of atomic structures. Indeed, as we have shown even for simple quantitative relations, such explanations do not lead any further and are, thus, superfluous and often even confusing.

Often we overlook the fact that it is very possible to accept the existence of ultimate small particles, but to refuse to use them as a basis for explanations. Such an explanation is based on an underestimation of the phenomena of nature. It is inadequate because it doesn't really explain the phenomena.

By no means does this chapter mean to be an attack, but rather a defense. Even as small as they are, the atoms have come too much to the foreground of human thinking. They must be put back in the place where they belong.

If we want to develop a chemistry that does justice to the fact that there are living beings, then we should see behind every phenomenon their "authentic wholeness," and deduce this wholeness from the phenomena. That also has to be done in a similar way with smallest particles.

Rudolf Steiner has indicated that we should be attentive to the great contrast between central and peripheral forces when studying natural phenomena. The central forces are well enough known. They are those forces which both radiate out from a center and are also directed towards it (see *Light-course*). Mechanics and essentially all of physics are based on working with these forces. In the realm of inorganic processes we can progress quite far, even if we only direct our attention to these forces. However, taking the central forces exclusively into account we can only partly understand the phenomena in the sphere of life. We can only understand the phenomenon of life if we pay attention to the interaction of those forces which radiate from a center outwards and those which radiate inwards from the universe, and which are fundamentally different. These latter generally act taking the whole into account. To understand these, we need to develop totally new conceptions of space.[9]

The atomic theory, as advanced by Dalton, is an attempt to take the explanations based on central forces to the limits. If we only think in this manner, then we likewise train our consciousness to only take notice of the spatial. Later developments naturally brought a much greater diversity and overcame some of the rigidity of the initial atomic theory. However, the new notions of the atom have hardly produced anything that could be used to break the bond with space. In broad outline, these notions must, of course, be discussed in grade 11; however, they are also not sufficient to explain common natural phenomena.

Our task is to educate the students in such a way that they can keep their eyes open to all directions. It is hard to speak to them in a concrete manner about the peripheral forces. But both in the realm of mathematics and in chemistry, it is possible to give a basis for independent research. If we continuously stress the importance in chemistry of the totality and of the many well-ordered relations that exist between phenomena, then we open a kind of gate through which we may enter the land of the peripheral forces.[10]

Endnotes

[1] *Goethe's Natural-Scientific Writings*, edited by Rudolf Steiner, 1883 (Kürschners ed., 114 - 117, 1.2) Vol 4-I Introduction, par. 7, Paperback ed. of the Introductions by Rudolf Steiner Stuttgart 1962, p 233.
See also "Goethe Against Atomism: Chapter XVII.

[2] Van Meurs en Baudet, *Inleiding tot de scheikunde* (Introduction to Chemistry), part 6.

[3] In principal there is no difference between the first set of numbers/ratios and the second; the second set is easily recognized as a "tidied" up version of the first, in that whole numbers are used and the ratios have been expressed in the smallest

possible whole numbers. Upon analysis, any substance will give us a ratio of the products of analysis, and any ratio can be reduced to one containing the smallest whole numbers. By choosing hydrogen as a reference substance, we get a network of ratios which are identical with "atomic masses" with only minor modification. (It should also be noted that, as Julius points out, these ratios are based on the same method [Cannizaro] which can be found in any chemistry textbook prior to around 1970, for example Mellor's **Modern Inorganic Chemistry**, ref 75. For a more modern approach, see **Chemistry** ref.86).

Julius follows the traditional approach to derivation of atomic and molecular mass and what we now call the "mole" at the time of writing the expression was g-atom and g-molecule.

[4] Note that Julius uses water as a reference substance when calculating the relative density.

[5] Berzelius empirical table of atom weights 1814; Stanislao Cannizaro, 1860 by gas vapor densities.

[6] The substance H has been detected and thus is a physical substance, for example in the reaction of zinc and hydrochloric acid in a solution of Iron-III-chloride; alternatively, the hydrogen gas produced by the action of zinc on hydrochloric acid in a Kipp apparatus hasn't got the same property.

[7] Nowadays the concept mole is used for both gram-molecule and gram-atom

[8] Steiner R, **The Origins of Natural Science**, Dornach, Dec. 1922-Jan. 1923, GA 326, Lecture 5.

[9] Adams, G., Whicher, O., **The Plant between Sun and Earth**, Stourbridge 1952.

[10] The writing of this paragraph was made possible through the study of essays by Eugen Kolisko, which were published in 1922 as "Hypotheses free chemistry in the sense of spiritual science" (Hypothesenfreie Chemie im Sinne der Gesiteswissenschaft) in "Aenigmatiges aus Kunst und Wissenschaft", Stuttgart 1923, a summary of lectures given during the first Hochschulkurs in Dornach 1920. For those who know these revolutionary essays, it will be evident that my work differs in many aspects.

VI

The Process of Dissolving - Phenomena, Concepts, Laws

The Most Important Phenomena

To begin with, we'd like to focus on dissolving a very well known salt. For example, take a blue copper sulfate crystal and put it in a glass filled with slightly acidified water. The crystal is bounded by planes, which meet at certain angles in distinct linear edges. However, after a short time we see that the crystal is decreasing in size, while at the same time the most prominent edges and points are receding and are being rounded. The crystal becomes ever smaller, and in the end it completely disappears. Around the crystal the water has become ever more blue, and at the same time it has acquired a sharp taste. The colored water initially sinks to the bottom of the flask, where it forms a layer. Then, however, even when the flask is absolutely at rest, the color moves slowly up through the whole liquid. For a long time the color remains most intense at the bottom of the flask, all the while fading and ever more going to the surface. The sinking of the colored water is caused by gravity—the solution is denser than pure water. The propagation to the surface occurs without any outward influence. In such cases where two substances mix, without any outward force being applied, we speak of diffusion.

Here, as in the case of chemical reactions, separateness and impenetrability are overcome to a certain extent. In contrast to this, certain properties of the dissolving salt substance (color and taste) are transferred *unchanged* to the solution. The changes in solution, therefore, are less intense than in the case of a chemical reaction.

To investigate if the solvent is capable of taking up an indefinite amount of substance, we mix water with copper sulfate crystals, and as soon as there is nothing solid left, we add more crystals. We repeat this and eventually reach a limit where the water will not dissolve more any crystals. We say the solution has become *saturated*. This state is reached more easily if we hang the crystals to be dissolved just under the surface of the solute. This creates a circulation because of the sinking of the solution.

Heating that solution will let more salt dissolve. Careful experimentation shows that at any given temperature there is a definite relationship of maximum amount of salt that can be dissolved by a given amount of water. In contrast, if a warm, saturated solution is cooled, copper sulfate will be seen gradually crystallizing out. Comparing these phenomena with those observed during a chemical reaction, then it becomes very apparent that in this case only gradually changing relations emerge, that here there are no fixed weight relationships. However, if as an exception, a definite weight-relationship does occur as for saturated solutions, then it is, however, still dependent on temperature; changing the temperature changes this ratio.

At times it isn't so easy to distinguish a solution from a compound. Often it is just the presence or absence of a definite weight ratio that is utilized as a criterion to distinguish them.

If a piece of gold is placed in water, naturally it will not dissolve. However, if it is placed into mercury, then it *will* dissolve relatively easily. Evaporating the mercury by heating, then the gold is recovered as brilliant metal crystals. Also, a piece of sulphur is placed in water. Just as with gold, it also does not dissolve. If, however, it is mixed with carbon disulfide, then it *will* disappear quite quickly. Allowing the solution to stand then, the carbon disulfide will evaporate, and beautiful sulphur crystals will be obtained once again. (Note: Exercise proper precautions for TOXICITY HAZARDS for both Hg and CS_2, if these are demonstrated.)

The Concept of Solubility

It is nearly self evident that to define the concept of solubility, we have used those particular phenomena where a specific weight relationship shows up. Thus, we find the following phrasing: "The solubility of a substance (solute) in a particular solvent and at a particular temperature, is the *amount in grams* of this substance that will dissolve *in 100 g* of solvent *at that temperature* to give a saturated solution."

This definition is really quite problematic. Above all it would be more correct to start the sentence as follows: "We designate by the solubility of a substance...." Such wordings are really nothing more than looking for a way to express quite complex [qualitative] phenomena in a summary way using (quantitative) numbers. However, it would be better to state the definition as follows: "To express the solubility of a substance best in numbers, determine the maximum amount of substance in grams which can be taken up by 100 g of a particular solvent; each number thus determined is valid only at a specific temperature."

Not only is there this pure quantitative aspect, but also the qualitative one which had to be discussed first. The first thing that has to happen, naturally, is showing the phenomena, and then the related descriptive explanation. It is then possible not only to discover a certain regularity in the solubility or insolubility of certain substances in specific liquids, but this regularity can be understood and explained quite well.

The main rule for solubility is: the degree of solubility of a solid in a liquid or of two liquids in each other depends to a very large extent on their mutual kinship. In the sense used here, "kinship" means similarity, thus not the concept of chemical affinity which plays a role in chemical reactions and that often is based more on a certain antagonism of properties.

So we also find here a remarkable distinction between processes which occur when dissolving takes place and those which occur during a chemical reaction. This could be expressed as follows: In general there is little chemical affinity between substances that belong to the same category of substances.

There are mainly three groups of substances, within which the substances themselves are similar and where mutual solubility occurs.

1. Oxygen-poor combustible substances, such as petroleum products, fats and fatty oils, sulphur, phosphorus, carbon disulfide, ether, benzene, etc.

2. The metals. Most metals are soluble in mercury and molten metals.

3. Water and many oxygen-rich substances, most of which are chemically active or very susceptible to chemical change. In the first place, strong acids, bases, and their salts belong to this group.

One has to bear in mind that the rules in this area are not very hard and fast, and it is, therefore, very difficult to predict with any precision.

For example, there are many sorts of minerals with a high oxygen content, which chemically should be classified as salts; however, they are insoluble in water (ex: silicate minerals). In this case, the insolubility reflects an overriding, strong affinity with the earth element.

There are also numerous transition and intermediate substances, which mainly have a role in living organisms. The various sugars and plant acids come to mind. Alcohol, which is prepared from sugar, is both very combustible *and* very soluble.

Ether, which is prepared from alcohol by dehydration (which in this case corresponds to oxygen extraction), is very slightly soluble in water, consistent with our rule, but its combustibility has been highly increased. The very combustible oxygen-poor fats are insoluble in water, but in ether their solubility is good (as expected). Fats are slightly soluble in cold, but very soluble in hot alcohol. In every respect alcohol stands midway between water and ether.

We could start with the above qualitative treatment of dissolving, before treating the quantitative aspects. In an education that is not only concerned with teaching technical skills, but rather wants to open the way to reality and truth of the topics discussed, the student must be given the opportunity for forming many-sided impressions of the processes under discussion. They must be in a position to empathize with, even to connect themselves through their will to the processes. If they inwardly experience them in that way, they will create living inner pictures and concepts, which, however, will not as yet be sharply defined. Only later may we lead them into the barren one-sidedness of clearly defined abstractions.

If we follow this method, then those who are intellectually less capable will be able to follow the subject matter further than with the usual methods. On the other hand, the intellectually capable will be less likely to completely isolate themselves from the reality of nature, because they will not fall for the temptation to use only their head forces when building their picture of the world.

The Influence of Heat on Solubility

During solution solids change to a more mobile and less dense state. So something happens that also happens through the influence of heat. Based on these facts, we can understand why the solubility generally increases with rising temperature.

For gases it works differently. A gas will always be reabsorbed to a greater or lesser degree, when it stands over water. It goes from a more mobile and less earth-bound state to one which is denser. Its mobility is restricted and a greater binding to the earth comes about. This also holds as long as the water contains less of the gas per unit-volume than the surrounding air. In accordance with this, the solubility of gases *decreases* with increasing temperature. By increasing the pressure, i.e. by increasing the density, the amount of dissolved gas will increase.

Also, there a great difference with regard to the influence of heat on dissolving and on chemical reactions. Increasing the temperature will increase the speed of a chemical reaction—just as is the case with dissolving, but temperature doesn't have an influence on the reacting weight ratios, as long as it isn't extreme.

Phenomena of Osmosis

If we want to discuss osmosis and understand the state of dissolved substances more clearly, then we can begin best with the phenomena as they appear in the *animal* world.

While typical land animals such as mammals, birds, reptiles and arthropods (with the exception of most crab like animals) have a skin which is impermeable for water, nearly all other animals have a skin which is permeable for water. That can have remarkable consequences. On land a frog is constantly in danger of drying out. Placing a half-dried frog into water, it will quickly recover. Since a frog cannot drink, we could just as effectively have placed it with its stomach in a puddle of water or let water drop on its back. It is even capable of taking up sufficient water through only one finger so as to bring its water balance back in order. While in dry air, a frog is in danger of losing too much water, under water it continuously takes up water through its skin, and only through an active kidney excretion can it keep its equilibrium. A remarkable phenomenon is the following: While there is an active exchange between water and blood in a frog, the substances dissolved in the blood are retained. These substances are enclosed within the skin. Therefore, we speak of a ***semipermeable membrane*** and mean by that a skin that is permeable for water but not for the dissolved substances.

If a frog were placed in water of a higher salt concentration than that of its blood, water would be removed from the body as if it were drying up. Salt from the surrounding water cannot penetrate; however, water from the body fluids streams outwards.

A similar but much more drastic process takes place when salt is sprinkled on a snail or slug in the garden to kill them. It shrivels up completely.

Fish, such as salmon and eels, which swim from fresh to salt water and back, can only move very slowly through the intermediate brackish water, because they need time to adapt. Salt water fish suddenly exposed to fresh water will take up too much water; fresh water fish, placed in salt water, will be sucked dry as it were. If we know these things, then we can be extraordinary respectful of the resilience of many organisms of the sea shore, living between high and low tide. Seaweeds, for example, must not only be capable of withstanding the mechanical force of the surf, but must also be able to cope with drying out under the burning sun, and also with resisting osmotic swelling due to low tide and fresh rain water.

Amongst the well-known fish, there is one, the three-pronged stickleback (Gasterosteus aculeatus), which can be repeatedly moved from fresh to salt water without any harm being done.

It is really quite selfevident that our body keeps the concentration of the substances dissolved in our blood at a constant level. If the concentration became too high, then the red blood cells, for example, would shrivel. If the concentration were to decrease, then they would swell and even burst open. The tissues would also be damaged. It should be clear from the this that the concentration of the blood during blood transfusions is of the utmost importance. Our kidneys regulate the osmosis of the blood in an excellent way. In all the processes they perform—diluting or concentrating the blood as a solution, and what ever else takes place—they counteract the effect of osmosis. When discussing this topic, we should never forget to draw attention to the fact that processes in living organisms cannot unconditionally be subject to osmosis, because they themselves must first build up the osmotic systems. The fact that certain fish can adapt points in the same direction.

Explanation of Osmosis

To understand the phenomenon of osmosis in the living organism, we look to what has been simulated in the laboratory. For example, we can use a fresh pig bladder, filled with quite a concentrated solution of a colored salt, and hang the bladder in a basin with tap water. Then, by the spreading of the color we will see that the dissolved substance penetrates the bladder to some extent. Moreover, the solution will slowly dribble out of the opening of the bladder, even when the opening is held somewhat above the water level. Placing a glass tube in this opening, and making the seal watertight, we see that the solution rises up the tube. From this observation it follows that the wall of the bladder isn't quite (or wholly) semipermeable, but that there is a remarkable flow from the surrounding water into the solution. Water and solution stream towards each other—the emphasis, however, lies here with the water flow *into* the solution.

Having reached this point, the question arose if it was possible to make synthetic membranes which would be absolutely semipermeable and at the same time could withstand high pressures. This was accomplished for the first time by the great botanist Pfeffer.[1] He used cylindrical, porous earthen pots and filled these to the brim with a solution of potassium ferrocyanide and potassium hexa-cyanoferrate II— $K_4(Fe(CN)_6)$. Then he submerged it in a solution of copper sulfate. Where the two solutions met at the wall of the pot, a precipitate formed of cupric ferrocyanide [copper-II-cyanoferrate(II)]. This was indeed permeable to water, but absolutely impermeable for certain dissolved substances. With his experiments, Pfeffer wanted to imitate the plant cell in which the soft protoplasm membrane is supported by a strong cell wall. To further complete the simulation, he only had to fill the "cell," with a solution and carefully seal it. By attaching a pressure-gauge to the "cell" he was able to express the results in numbers. When such a cell, filled with a more or less concentrated solution, is placed in pure water, nothing of the dissolved substance goes out, while the water flows towards the dissolved substance, thus entering the cell. This phenomenon is called osmosis. The cell exercises an osmotic suction on the surrounding water. As the pressure increases, the inflow of water decreases, till this stops at a certain maximum pressure. This maximum pressure is proportional to the concentration of the solution. This pressure is called the "osmotic pressure" of a solution, even when this solution is in a glass vessel and thus cannot exert a pressure. Of course, we are talking about the pressure which occurs when the cell is surrounded by pure water. If the maximum pressure is exceeded, then water will flow out through the wall of the cell and the solution will become *more* concentrated. Normally, if the (dialysis) "cell" filled with a solution is placed into another solution, then the liquid will move towards the solution with the highest concentration.

The osmotic pressure of a solution, as it can occur in certain plants, can be very high. Pressures of more than 100 atm. have been measured. In such cases we are dealing with solutions which simultaneously contain different dissolved substances. However, we should remember that then we are dealing with the highest possible pressure. In nature, the conditions for such pressures to actually occur will hardly ever be achieved. This is because there isn't enough water, or because the outside liquid is a solution with a certain concentration and not pure water. A lot of these phenomena can be understood if it is realized that the state and the movement of a dissolved substance in a solvent shows a strong analogy to the situation with a gas. With both situations, the tendency of unlimited expansion, and that of unlimited dilution is found. If the volume in which a gas or a dissolved substance can expand is limited, then both will aim to fill the volume uniformly.

A great difference will be found in the speed of diffusion. In the case of solutions this is much smaller than with gases. However, the fundamental difference is that a gas can diffuse through empty space, whereas a solute can diffuse only in a container filled with fluid.

If it were possible to fill a semipermeable membrane that was indefinitely expandable with a solution and then set this in pure water, then this membrane would expand in the same way as does a gas-filled balloon in a vacuum. In both cases, the membrane is pressed outwards. In both cases, resisting this expansion results in an analogous pressure. Only osmotic suction with solutions does not have an analogous phenomenon with gases. But, also, with gases we are dealing with an expansion of the solute in the solvent, and thus with a diluting of the solute.

The Concept of Concentration

We have come so far that the concept of "concentration," which we have used already, can now be given a precise meaning. By agreement, we express the concentration of a substance in a solution (or in a gas) numerically as the number of moles of the substance in a volume of 1 liter. Or, more briefly, the concentration is expressed chemically in moles per liter (physically, concentration is usually expressed as g solute/100g solvent).

Initially this will very likely surprise the student, and they might probably experience some difficulty because they previously learned that to express the solubility, they investigated how many grams of substance dissolved in 100 g of a solvent.

There are very good reasons for making these agreements in this form, and it is important for the student that these be explained clearly. We will shortly see that it is better to follow another path to express the concentration in a way more in line with reality.

As we have seen, solubility is related to the specific properties of two substances and their mutual relationship. The definition must, therefore, express something about the possibility of expressing the quantitative relationship precisely and unambiguously. That can be done best by using weights.

The "concentration" of a dissolved substance relates to its varying presence at a certain specific place. The nature of the solvent is unimportant, because the dissolved substance can be present in varying concentrations in the same solvent. The solvent only functions as a certain space in which the solute can expand. Actually, it is only a question of the degree of filling a certain space—thus, of the relation of a substance to space as such, and not of the relation of one substance to the other. If we were only interested in the variations of concentration of one single solute, then it would be absolutely sufficient to express the magnitude of the concentration in *grams* per liter. However, concentrations of different substances must often be compared, and then, as will now be discussed, the results are clearer when the concentration is expressed in *moles* per liter.

Comparing the solutions of various substances in a specific solvent—for example, water—resulted in the following rule:

If, we dissolve *various* substances in one specific solvent, in such amounts that the magnitudes of the osmotic pressures are equal, then the magnitudes of the freezing point lowering, and the boiling point elevation, and the depression of the vapor pressure *are all equal*.

Also, for the *same* substance at different concentrations in the same solvent, the magnitudes of osmotic pressure, the lowering of freezing point, the lowering of vapor pressure, and the elevation of the boiling point, are all *proportional to the concentration* of that substance (solute).

Note that such rules are only accurate when the solutions are quite dilute.

It has further been found that the magnitude of the osmotic pressure is *not* dependent on the nature of the solvent, but that this is not so for the three other phenomena. Thus, the osmotic pressure will be equal if the same amount of a substance is dissolved in different solvents, all making 1 liter solutions; but, the magnitudes of the lowering of the freezing points will be different. This last is really self-evident, because the lowering of the freezing point is a change of a property of the solvent caused by the solute, while the osmotic pressure is an expression of the relation of the solute to the volume. Of course, the osmotic pressure must always be measured at the same temperature to eliminate the influence of heat, which is considerable.

Thus, for solutions of the *same* substance in different solvents, the osmotic pressures are the same, so the [osmotic] concentrations are equal.

What is the case if solutions of different substances are compared, which have equal osmotic pressure? In this case the amounts of dissolved substance are in the same proportion as the molecular weights.

By agreement, also in this case, we say that the different substances have the same concentration. If different solutions at a specific temperature have the same osmotic pressure, then the concentrations of these solutions are said to be equal.

When equal amounts by weight of various substances are dissolved in equal volumes of the same solvent, then the ratio of the osmotic pressure is inversely proportional to that of the molecular weights. The value of the osmotic pressure, and, therefore, that of the concentration, is proportional to the weight of the dissolved substance and inversely proportional to the molecular weights.

$$\text{concentration A : concentration B} = \frac{W_A}{a} : \frac{W_B}{b}$$

W_A = weight of A $\qquad$ a = molecular weight of A
W_B = weight of B $\qquad$ b = molecular weight of B

If we want to express the amount of substance in gram moles, then the weight of the substance in grams (W_A) must be divided by the gram-molecular weight (a).

$$W_A \text{ grams of A} = \frac{W_A}{a} \text{ "gram moles" of A}$$

Substituting this into the first equation, results in:

conc. A : conc. B = quantity in moles of A : quantity in moles of B or, in other words: If we agree to always express the concentration in moles (moles/liter), then we have met the demand to be able to compare the concentration of various substances in a simple and clear manner.

As we expected, we can use the reasoning outlined above for the concept of (osmotic) concentration in the same way for gases. When the gas pressure and the osmotic pressure are the same for a gas and for a dissolved substance, then the amount of gas (or concentration expressed in mole/l) and of the dissolved substance, expressed in moles, are the same.

It is always difficult to keep in mind which is the primary concept and which is the derived one. The concept of concentration has been defined with the aid of the concept of the mole, which itself has been derived from the concept of the molecular weight.

In developing the concept of the molecular weight in the last chapter, we started out with the idea of vapor density of the substance in the gas phase. We get the value of the vapor density by comparing the weights, of equal volumes of various gases, at the same pressure. To develop the concepts of vapor density and molecular weight, we had to start with substances at the same concentration. In the concept of the molecular weight, used to define the concept of concentration, the concept concentration is already contained.

This implies nothing more or less than that to reach a perfect definition of the concept of concentration, we have to start off with the pressure of a gas or the osmotic pressure.

When different gases and different dissolved substances cause the same pressure at the same temperature, then their concentrations are equal.

To determine the concentration of a substance, it seems appropriate to measure the pressure. However, because the measurement of the osmotic pressure is quite time consuming, it is replaced by the determination of the lowering of the freezing point. If the value of the specific (molar) freezing point depression for that solution is known, it is then quite straight forward to come to a conclusion about the magnitude of the osmotic pressure.

I can very well imagine that someone will object: these thoughts are too difficult to be discussed with the students, in this form. I am actually of the opinion that this is correct. However, if these concepts are introduced, but the students are allowed to use them without explaining their basis, then this makes it impossible for them to find their own way independently into this topic of chemistry. It would then be better not to discuss these concepts in school at all.

If, however, you want to discuss the concept of concentration then, in my opinion, it is best to start by defining it with the help of the concept of pressure, as has been done here. Then, it can be shown that the concentration of a substance is proportional to the magnitude of the weight of the substance in a certain volume. And furthermore, for different substances having the same concentration, it can be shown that the weight (mass) is proportional to the molecular weight (molar mass).

In summary, we may say that concentration can best be specified by using the pressure; if, however one wants to express the concentration in (chemical) units of substance, then using moles is the best way.

The Concept of Dissociation

In our discussion thus far, we haven't acknowledged the fact that when salts are dissolved, for example, it gives rise to a kind of "dissociation." Initially, this is shown by such solutions exhibiting an osmotic pressure, freezing point depression, etc., which is *greater* than what would be expected from the molar concentration (in moles/liter), or, in other words, greater than what would be deduced from the magnitude of the molecular weight. If a substance at sufficient dilutions causes a freezing point depression which is twice as large as normal, then we must assume that during the process of solution, the substance is split in two. In many cases, the depression will be less than twice the amount expected. We then imagine that the substance is only partially dissociated.

Again and again, we can be enthusiastic about scientific thinking, because it always manages to put into precise concepts even such vague intermediate states, and, therefore, they become accessible to manipulation.

However, we have grown accustomed to utilize them with the aid of "particles," "molecules" and their "dissociation products." Because we don't know how many particles there are, we assign this to the variable n for example. That means nothing more or less than that one isn't making use of particles. Here, also, we are only concerned with ratios of numbers and amounts.

The following definition for the degree of dissociation can be found in an otherwise excellent high school textbook:

"The degree of dissociation of a substance is given by a ratio expressing the number of particles that have dissociated, to the initial number of particles."

From a phenomenological perspective the definition should be: "The degree of dissociation of a substance is given by a ratio which expresses the amount of substance that has dissociated, compared to the initial amount of substance."

Every logically thinking student should be offended by the way such definitions are stated. In many cases initially we have two substances which partially associate into one. Remarkably, by "initially present molecules" often something else is meant than what was intended. In a systematic compendium of the sciences[3] we find: "By degree of dissociation one means the part of the total substance that is present in a dissociated form." Here they don't speak about smallest particles and also don't use the incorrect concept "initial molecules." As far as it goes, this definition is better than the others. The ideal solution most probably is to use a combination of both definitions: "The ratio of the dissociated or associated amount of substance is called the degree of dissociation." This ratio is expressed as a fraction.

Continuing on our chosen path, we could show that in nearly all instances where we believed we had to take recourse to thinking of smallest particles, this wasn't necessary in actual fact. For example, we introduced the weight of an atom of hydrogen in a certain calculation, only to eliminate it in the next step. In most cases, we are *looking for support* for our imagination when we use the concept of "smallest particles." So, when we come to the point where quantitative ratios appear, then we frequently try to avoid agile thinking, which is just what is absolutely necessary here.

Endnotes

[1] Pfeffer, W. *Osmotische Untersuchungen*, W. Engelmann, Leipzig 1877.

[2] " $\rightarrow$ " indicates "equivalent to."

[3] Algemeenrepertorium, Elsevier, Amsterdam-Brussels, 1955.

VII

The Great Matter Cycles: Chemistry in the 11th Grade

Usefulness of the Element Concept

In grade 11, in the way we present and consider the various substances, we must bring in the greatest contrasts—yet in such a way that an unfragmented image of the world arises. Certainly, we should characterize the traits and significance of each individual chemical element. Our starting point, however, should be quite different than the conventional concept of an element. If, nevertheless, we build our class on this conventional element concept, we will present the students with a problematic historical lineage and even carry it further into building up an image of the world for ourselves which is dry and rigid.

What has modern chemistry emphasized most? The Dutch word "scheikunde" says it all: literally "the art of dividing," i.e. analytic chemistry. We take the available substances and analyze them. In this way, people hoped to gain insights into their nature, how they might have arisen, and then to "take the bridle in their own hands" and fashion substances at will for themselves. Thus, they found that some substances could be taken apart by chemical means, so that two or more products arose; but they also found other substances which could not be broken down further into "components" by chemical means. These were termed chemical "elements." Implicit in this term is the conviction that "elements" are the "building blocks" of the whole house of nature; those substances which could be analyzed were termed "compounds." Here again, at the root of this term, lies the idea that the essential nature of substance would be fully explicable as soon as we knew how they arose out of the (constituent) "elements".

A powerfully suggestive influence arises from such tendencies in thinking, so that it is very difficult for a person schooled in natural science to imagine that, e.g., carbon dioxide (carbonic acid gas) is a unity in the same sense as oxygen is.[1] Yet, both gases occur in nature, each in a characteristic way. We don't usually notice that one substance is an "elemental substance" and the other a "compound."

Without our being conscious of it, this characteristic tendency involves a certain degree of anthropomorphizing. If we as humans build something, we must select our raw materials from that which nature makes available to us, and then we join them together according to a plan which we carry within us; the materials come from outside nature and the plan from our inner world. But, when nature builds, we cannot speak of materials assembled from outside, nor can we speak of a plan projected out from some inner realm. The capacity to build and the plan itself, both lie within the same plane.[2] Both the characteristics of the materials and the things which arise as an expression belong together to the same essence. With nature, we are simply dealing with one great, unified whole, to which each part is oriented and also—as is true for every being—each part related to every other part. Whether we have an elemental substance or a compound, we must treat both as differentiations of the whole and not as building blocks, in an anthropomorphic way.

The Periodic System of Elements in Chemistry Classes

One very real and significant result of the research done in the 19th century was the discovery that all elements were members of *a definite system*, which was termed the "periodic system of the elements." This system described the underlying natural order in such a clear way that it was even used successfully to predict the characteristics of elements as yet undiscovered. The periodic system describes the characteristics of each element family, and how it is deeply linked in a relationship to all other elements. The fact that these characteristics are so clearly interrelated actually arises from the fact that they are actually part of an all-inclusive whole. So, it is clear that we have to trace the individual substances back to this whole, not derive the whole from the sum of individual substances. Although we are dealing here with an impressive architecture of the whole, nonetheless it is difficult to build up a living image of the world for the students if we use this system as the starting point. If we did this, it would be necessary to erect boundaries to knowledge in order to protect them—which is by no means possible in school; and to explore outside these limits would require capacities which are only acquired as a grown-up, through years of study and schooling.

The way the conventional periodic system is presented in textbooks is, therefore, excessively dull and abstract. Many museums make the attempt to present it in a more interesting and understandable way (usually by showing somewhat larger quantities, each in a flask above a nameplate). However good this may be, it is actually a symptom of the tendency of modern people to permit the wholeness of the world to be dissected into isolated pieces.

However, other approaches to the periodic system of elements are possible, which allow the impressive unity and the pure, harmonic relationships to receive their due (see Bindel's work[3]). However, even when this succeeds, there are two great difficulties bound up with using this as a starting point for describing a living world. First, it is hard to show the connection with the human organism from this viewpoint; second, this system is not the product of an imagistic, picture-thinking. If, however, a person does use such a picture-thinking, they can discover totally different lawful relationships between the elements, which are significantly more understandable and useful in teaching.

That doesn't mean we shouldn't study the periodic system. But, there is a tremendous difference whether one takes it as a starting point, or else describes it as a significant, noteworthy discovery—though not the key to all substance.

The Human Organism as a Key to a New System

The first requisite for teaching is: strive so that the student receives a living and comprehensive picture of the world and of the human being. Often, this will only be possible if we choose a somewhat unusual path. Naturally, as the students grow older, we should not ignore the fact that it becomes increasingly important to make them familiar with conventional methods, so that students are familiar working with them and even practice them in class.

Where do we find the key to such a unified image of natural substances? How do we build up an image which is exact in the modern sense, but which also has not destroyed the connections to life and—most importantly—in which the human being receives his proper place? The answer is just as surprising as it is simple: from the analysis of the human body.

It was through an analysis of the substances of earth's "body" that chemists discovered a periodic system embracing all the elements. So, if we analyze the human body we would expect to find just those elemental substances which are especially significant for the life-processes. We might even expect to find no other groupings or characteristic laws than we have already met.

Since in teaching we wish to remain as much as possible in the realm of true and living concepts which can be developed by our understanding, we will not start by filling out a list of elements, but rather in such a way as to give a picture of them, which must happens first, before such a table can arise.

First, we put in front of us the wondrous whole of the living body, where each part stands in a developing, living connection or relationship with the whole, and with every other part. This image is presented in contrast to the corpse: here everything is rigid and frozen, nothing holds the whole together, and all portions are in the process of falling apart. Then we consider how the chemistry that occurs after death carries this process much further. A person only has to consider these decomposition processes in thought in order to get a strong impression of the activity of the life processes.

Now we enumerate the chemical elements which are the most important products of decomposition or analysis of the body. In teaching, we can introduce the names and the symbols as an abbreviation.

Elements Commonly Found in the Human Body[4]

Traditional Name	Chemical Name	Chemical Symbol	Wt. %$^{\pm}$
Water-gas	Hydrogen	H	10.0
—	Nitrogen	N	3.0
Acid-fomer	Oxygen	O	65.0
Sun bearer	Sulfur	S	0.25
Light bearer	Phosphorous	P	1.0
—	Carbon	C	18.0
Natrium	Sodium	Na	0.15
Kalium	Potassium	K	0.35
—	Magnesium	Mg	0.05
Lime	Calcium	Ca	2.0
Silica	Silicon	Si	trace
Ferrum	Iron	Fe	0.004
Fluor	Fluorite	F	trace
—	Chlorine	Cl	0.15
—	Iodine	I	trace

So far, this is hardly a comprehensive whole. However, there are certain points from which striking classifications appear.

There are three substances which distinguish themselves from the others by their mutual relationship and their characteristic behavior with regards to oxygen. Most elements react with oxygen, and correspondingly become activated. Substances which are not attacked are on the whole in a passive state or more or less chemically inert. Fluorine, chlorine, and iodine, however, which together with bromine are called the four halogens, have hardly any affinity for oxygen, notwithstanding that they are chemically very active. We will begin by taking these substances out of the relationship with the others.[5] Then there usually remains twelve substances.

We now consider these substances from the point of view of the four elemental qualities—Earth, Water, Air, and Fire. Then, we find something very remarkable: hydrogen, oxygen and nitrogen are gaseous; they are typical components of the atmosphere and, thus, related to the *Air* element. Carbon participates in the form of carbon dioxide, also as an oxide. Sodium, potassium, magnesium and calcium together form the basis for the salts of the sea; they have a special connection with *Water*. Silicon and calcium are typical contributors to the crust of the earth—they are, therefore, related to the *Earth* element. Although iron is also part of this group, it belongs in a different arrangement, as we shall see later. Magnesium is also an important constituent of the earth's crust and so resembles calcium. However, as we will show later, it is caught up in a different and contrasting play of forces. We have already repeatedly come to know carbon, phosphorus, and sulfur as substances with a special connection to *Fire*. We can more or less include hydrogen here. All this and much more is expressed in the following diagram.

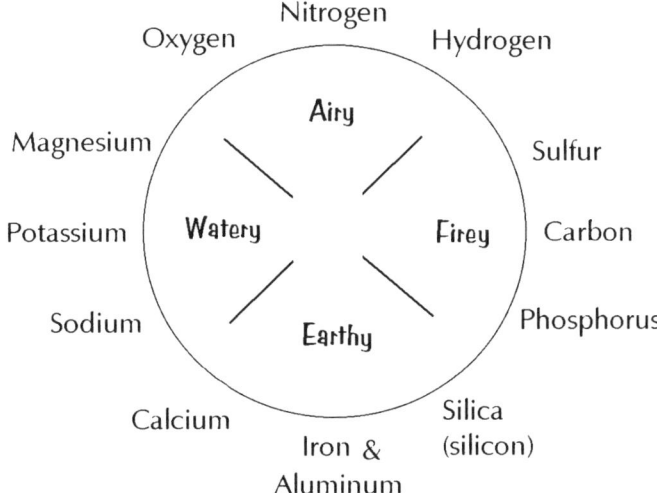

As we shall see, this circle contains mainly those elements which are the dominant constituents of the crust of the earth, the oceans, and the atmosphere, because of their weight or their volume (e.g., nitrogen). One striking exception to this pattern is aluminum. It is one of the most important rock forming materials, and although it belongs in this system, for reasons that will be explained further on it isn't found in the human body. Iron is the only heavy metal that is found in substantial amounts in our body.

At a later stage we will consider whether the heavy metals form a separate group, with their own laws. They do not belong to those substances which play a large part in the formation of the crust of the earth. They can be detected in the human body, but only in trace in amounts. Based on its overall properties, iron belongs in the heavy metal group, but in respect to the amounts found in the crust of the earth and in the human body, however, it belongs in our elemental circle.

In our discussion we have treated the elemental circle *as if* it were an expression of a strict rule of nature, and this is really so, as we shall see. The properties of the various substances are strictly related to their position in this system—in other words they correspond to the position of each within this diagram.

The Twelve Substances as a Representation of the Realms of Nature

We wish now to penetrate more closely into the laws which determine the connections between the characteristics of these substances and the whole of nature.

We can select a simple experiment as a starting point. We take a piece of sodium and leave it uncovered in a humid atmosphere. We see how it slowly disappears and in its place is a watery-fluid residue. If we bring this fact in relation to the fact that sodium is extracted from the most prevalent sea salt (halite), then we have an excellent picture of a general rule, which particularly applies to the substances of this circle. This rule may be expressed roughly as follows: *The characteristic properties or traits of a substance from our elemental circle may be deduced from the quality of the realm of nature from which it arises.* If one of these substances is active in this or that way, then that shows its inclination and ability to bring forth a miniature copy of that specific realm of nature. We can think: our piece of sodium was trying to re-create a miniature ocean, as it were.

We can get a still more concrete idea of the significance of the elemental circle as we investigate the characteristic realm of nature for each of the substances in the circle. This can only be briefly indicated here; when we discuss the individual substances later on, we will go into this at length.

nitrogen	=	the most important atmospheric-forming gas.
oxygen	=	gas of the air, constantly reactivated by the sun.
hydrogen	=	gas which strives intensely towards the periphery.

magnesium	=	substance which forms the kernel of chlorophyll – where the earth opens itself up to the sun.
sodium	=	the most important basis for the salts of the sea. Particularly absorbed by humans and animals. Most important basis for the salt in blood.
potassium	=	the most important basis for the salts absorbed by the plants.
calcium	=	the basis for chalk. This substance continuously moves between the solid and liquid state.
aluminum	=	important for rock formation, basis for clay.
silicon	=	the main component of the crust of the earth, important for rock and crystal formation .
sulfur	=	the substance which occurs in elemental form especially in association with volcanoes.
carbon	=	the most important substance for the synthesis of materials for the living organisms, especially the plants.
phosphorus	=	presents images of the starry heavens.

If we place these facts in our elemental circle, then a remarkable image of the relationship of chemical phenomena with nature results.

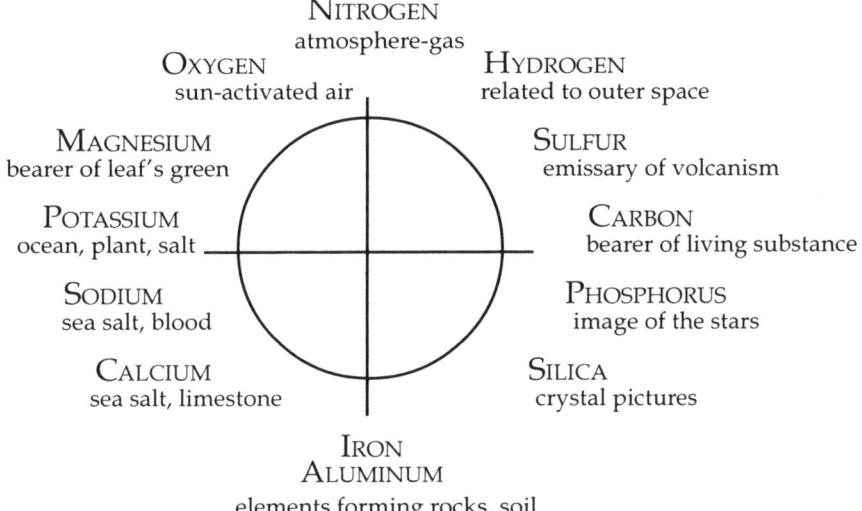

If we make a drawing of what was initially described with words, something surprising emerges.

Insofar as we dissect the substance of the human body, we get substances which together represent the totality of nature.

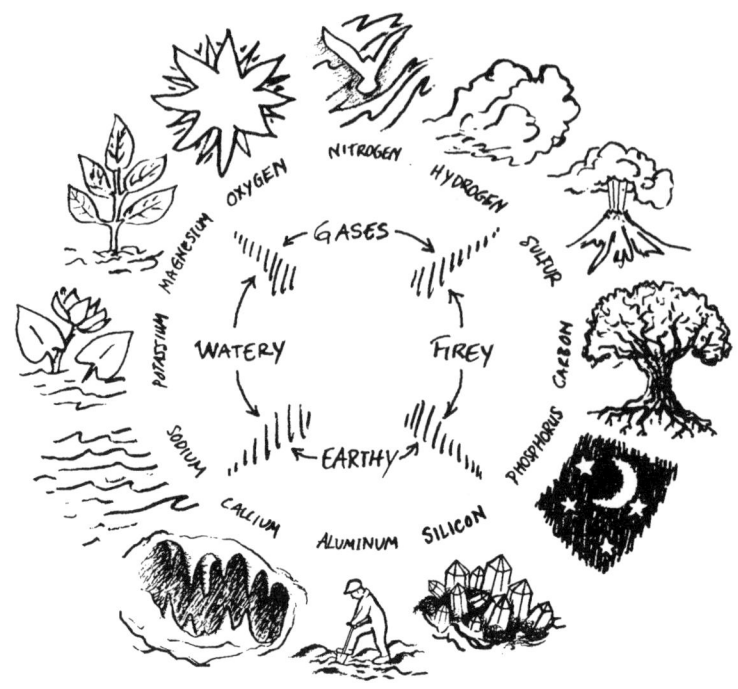

What has been expressed from time immemorial through mythology emerges here as the result of science: *the world may be compared to a disintegrated human—the human is a summary of the world. The relation of the human to the world is as that of the microcosm to the macrocosm.*

We now have the possibility to deduce in an exact way the particulars of chemistry from the vast totality of nature.

However, we should remember that the laws we will find and use can only be utilized with the aid of an intuitive-pictorial thinking.

Usually the properties of a substance are discussed as if they were unrelated. In reality the manifold properties always form a totality. This wholeness has a certain structure; we could even speak of an overall form which correlates to the above rule. Furthermore, the manifold properties of this group of elements form a still larger, highly-structured whole, wherein the structure of nature reflects itself. *Thus we will only understand the properties of the elements we are discussing, if we bring them in relation to the whole of nature and to the human body.* We will not seek the source of particular properties in the smallest parts of a substance, but in its position and task in the whole of nature, in which it is only a part. We are thereby not asserting that there is no relationship between the properties and the atomistic structure of the substance, but rather, that this isn't the cause, but at most is an effect of the place and task of the substance in the whole.

Indications for Teaching

Among the most important tasks of teaching is the *striving to give the students a powerful image of the all-encompassing tasks of matter, tasks which serve the world and serve the human being*. Each of the elemental substances in the elemental circle is tied into a mighty movement or *transformation which follows typical paths* for each elemental substance. We can speak of a "circulation." With this treatment, we come into contact with something quite living. Based on my experience, this aspect of the subject matter is especially suited to allow the students to write independent compositions after the class discussions.

But clearly, we also should not neglect the aspect of chemistry where the fundamental/elemental-weight is based on the exact course of certain reaction processes, and on the association and formation of certain substances. Thereby a person will also pay careful attention to the formulas and chemical equations. These things ought to be pleasantly dictated, since they are often very abstract and require exact formulation.

For the purely theoretical things, such as *the basic work with formulas and chemical equations*, the quantitative relationships and computations based on formulas could be laid out in a special book.

Next, we pass over to a detailed description of the elemental substances. This will also illustrate the basic points which we have just presented.

Oxygen

Oxygen as Matter Awakened by the Sun

If we want to acquaint ourselves with oxygen in its most active state, we must turn to the atmosphere, the sea of air around the earth. Oxygen behaves, on the one hand, in quite the same way as the other atmospheric gasses: it is transparent, odorless, without taste, in short quite invisible. On the other hand, it rushes into many processes to a greater extent than the other atmospheric gasses, whereby it ceases to be a gas. However, it also is continuously being set free from its compounds and always reactivated. This occurs under influence of sunlight during photosynthesis in the chlorophyll. We have thereby pointed out a major characteristic of oxygen: through the influence of sunlight it is transformed into a chemical element and as such is activated.

Oxygen as Atmospheric Gas

Not many substances in nature occur as chemical elements. Most of those are, moreover, totally or to a large degree chemically inactive.

Oxygen occurs in the atmosphere as a chemical element, though it is the most chemically active substance.

Furthermore, oxygen is remarkable because it is a gaseous element. Most elements are solid, only two are liquids (at room temperatures), and only a small group is gaseous. That suggests an important property of oxygen: its mobility.

The atmospheric gasses rank among the most rarefied and mobile substances we can find in nature. They participate in the most delicate manner in the changing relation of the earth to extra-terrestrial space, especially to the sun. They continuously whirl with the wind, they rise and fall, they expand and contract, they release moisture or absorb it, and they are heated up and cooled down. They are also incorporated in extraordinary phenomena of cosmic radiation, radioactivity, ionization, etc.

By confining the gas in a closed space, as happens all too often with physical experiments, very little remains of its active, natural, light-penetrated state.

We have to consider all this when we want to form an image of oxygen as part of the atmosphere. Otherwise, we are in danger of producing a pale image of such a substance.

Furthermore, it is very typical of the atmospheric gasses to be present everywhere. Only solids exclude air from their space; natural waters always contain some dissolved air. Of course, nitrogen is much more omnipresent than oxygen due to its higher concentration in the atmosphere.

We could call oxygen the "omni-active" gas. It hardly ever is absent near the surface of the earth, when a living organism wants to breathe, or when we wish to light a fire.

Oxygen as a Messenger from the Sun

It is primarily due to the activity of oxygen that we find in nature a perpetual unrest and tension, which results in numerous processes. Specifically, the bodily functions of the living beings are stimulated and kept going by breathing in oxygen. If we consider the manifold occurrences that are caused by the inhalation of air as one large earth-embracing process, than we must take note of its incredible extent. Immense amounts of substance are continuously being displaced and transformed through the forces of life. Through a lot of painstaking research it has been shown that, even going quite deep down into the earth's crust, a substantial amount of the substances are caught up in immense circulation processes, whose driving force depends in part upon living beings.

However, not only does oxygen set matter in motion through life, it also does this directly through chemistry. The fact that fires are always breaking out, that metals rust to ores (oxides), is primarily caused by oxygen. Such processes also bring substances into circulation again.

Everyone knows that these processes only take place, because oxygen is continuously being set free and activated by the sun. Rarely, however, do we consider that everything that starts from oxygen is an extension of the activity of the sun. And yet this may be easily understood by the following argument. Imagine we blotted out the activity of the sun and only the earth-forces remained, then everything around us would appear rigid, frozen, and gloomy. In reality the atmosphere is filled by the sun with light and warmth. That again calls forth a powerful circulation of air, water vapor, and water. A major activity of the sun is awakening the earth from dismal rigidity and isolation.

Oxygen does something similar with the substances it attacks. Often light and warmth are emitted, but primarily it calls forth activity, interaction, and mobility, as we'll see in detail later on.

Not only does the free oxygen diffuse in the regions illuminated and moved by the sun, but it also descends continuously as a sun messenger into the depth of matter, and brings the sun's enlivening influence there.

The astonishing thing is, however, that not only does oxygen carry the activity of the sun further, but it also intercepts and partially halts and regulates this influence. In the higher layers of the atmosphere oxygen acts as a chemical protective shield or filter and reduces the excess of this influence. Thus, the ultraviolet radiation, which causes our skin to darken, for example, remains in acceptable boundaries. Oxygen is thereby changed by the light, which it consumes, into a somewhat more dense, highly active state. This form of oxygen is called ozone. From the small amounts of ozone in the atmosphere emanates a purifying, healing influence, which is particularly noticeable in spring.

Thus, oxygen is activated by the sun in two steps: first through the process of photosynthesis and second through the formation of ozone.

Oxygen and the Crust of the Earth

If we were able to completely decompose the crust of the earth, we would find a lot of oxygen, even more than is found in the atmosphere.

If we calculate the weight of oxygen in the crust of the earth, we find that weight is about as large as that of all other elements together. More or less complicated oxygen compounds form by far the largest part of the crust of the earth.

Thus, in this realm a fair number of substances have largely come to rest and solidification. In a manner of speaking they are burnt out. In the mineral state oxygen, therefore, is completely at rest. We can speak of the end point of the oxygen process. This applies especially to quartz and siliceous substances.

Therefore, in the crust of the earth immense amounts of oxygen have become a solidified mass of rocks or semi-precious crystals. The most active gas has been immersed in a realm which has for the most part been separated off from the interaction between earth and cosmos, from the large circulation processes. The restless carrier of life has come to rest, in a realm where death reigns.

We have here a beautiful example of the dramatic events that take place in the world of substances, or that are impressed therein as memorials of a titanic past. The oxygen processes take place between two extremes. In the beginning they are similar to the activity of the sun, which always brings about renewal, change and mobility in the environment of the earth. At the end they are caught in a sphere that is totally alienated from the sun's activity and where all transformation or change has ceased. While renewal always points to the future, the rigid mineral world shows traces of a more flexible past.

Oxygen is attuned to the immense contrast between sun-connected life and earth-connected death.

In this context it is remarkable that the pure, oxygen-rich quartz crystals will let light pass into its well-formed inner space to a degree no other crystal will. Even in this earthbound state oxygen hasn't quite lost its relation to the sun.

Oxygen and the Hydrosphere

If we analyze water we find oxygen again as the main constituent of the products of analysis. In the production of water from hydrogen and oxygen, oxygen reaches a just as relaxed state as with the formation of minerals; however, no rigidity and isolation occurs.

It is typical for water, that it encourages every manner of interaction, through its ability to dissolve substances, through its mobility and tendency to circulate.

What are the continuous motions of waves and spray, other than an interaction with the air?

Not even the minerals are left at peace. They are washed over by water; in every crack that permits entry, water surges in, causing the mineral to lose some of its solid components, which are subsequently washed away by the water.

The absorption and giving off of heat plays a very great role as does also the stimulation of the exchange of heat world wide. We will go into this more fully when we discuss hydrogen. The interaction with light is also remarkable. Everyone is familiar with the play of the reflections, the sparkling of light and colors on a water surface. However, we take too little notice of the beautiful play of light which we can find in the top layers of the oceans. Here the crystalline transparency and the endless dance-like movement of the surface cooperate.

We can't really speak about the chemical activity of water itself. Yet, it is the preparer and unsurpassed carrier of the activity of other substances. It seems that chemical reactions can take place only when there is at least some trace of water present.

Water seems to have a special task when it comes to breaking boundaries and combining things, which are separated and lie apart.

Life without water is absolutely unthinkable. If a plant lets its seeds ripen, then the chemical processes come to a standstill, as soon as the plant repels water. If we want to let a seed ripen, the first thing we need to do is add water.

We recognize most clearly the part oxygen plays in the nature of water in bringing about of interactions, and in how it enhances reactions from the realms of chemistry and of life.

Oxygen and the Tria Principia of the Alchemists

We have remarked before on the extraordinary ability of the alchemists to express magnificent overviews of the structure of nature in concise images and symbols. The order of the Tria Principia is very apparent in the case of oxygen.

In the crust of the earth oxygen is in the "Sal" state. Here we have a realm that once was mobile and capable of transformation, and that has slowly changed into a state of rigidity and even of crystallization.

In water oxygen is in the "Mercurial state." Here everything is movement and interaction.

In air, oxygen is active and often induces reactions which release heat. We can speak here of a "Sulfur state."

Even if these states wherein oxygen manifests itself are very different, certain principles remain: reactions, from the realms of chemistry and life, are always made possible or enhanced. By means of atmospheric oxygen such reactions are activated. Water provides the realm wherein the reactions can occur. The crust of the earth acts as foundation–it supplies material and mechanical support.

Further Effects of Oxygen in Chemistry

As a chemical element, carbon is extremely rigid. Through combustion it is converted into carbon dioxide ("carbonic gas"). Not only is this substance very mobile, but it also develops a many-sided chemical activity. Furthermore, it easily dissolves in water. When carbon is allowed to burn with a restricted supply of oxygen, carbon monoxide results. This is a mobile gas, but it isn't very active chemically and not very soluble in water. Moreover, it is very poisonous.

Sulfur is absolutely insoluble in water and isn't attacked by other substances to such an extent that it is found as a mineral. If sulfur is ignited, then a chemically active, rather soluble gas results: sulfur dioxide. If we oxidize it further, then sulfur trioxide is formed, which – together with water – forms the extremely powerful, chemically active sulfuric acid.

There is also a gaseous sulfur compound that occurs in quantity without oxygen: hydrogen sulfide. It is extremely poisonous and chemically not very active.

Phenomena as described above are symptomatic of the effect of oxygen. If a substance is attacked by oxygen, then the result is generally increased mobility, enhanced chemical activity, and a greater solubility in water. As a rule, poisonous effects decrease through the influence of oxygen. Thus, oxygen enhances the chemical activity of other substances and makes them more accessible to the forces of living systems. Furthermore, it brings them more into the realm of water, e.g., the realm of chemistry and life.

Oxygen draws substances out of their isolation and inserts them into the great matter cycles. It creates interactions everywhere.

However, this is not so evident in many cases. If we heat a clean piece of copper, then it will form a black layer of oxide. This is still not very active and is insoluble in water. However, if we let it interact with dilute sulfuric acid, then we see that although the copper isn't attacked, the oxide dissolves. Thus, an enhancement of activity has taken place.

In addition, we must take the following into account: as soon as we dissolve the combustion products of sulfur, carbon, and phosphorus in water, then we get acids, while the metal oxides act as alkalis (bases). Two opposing chemical activities thus come about as the result of the different characters of the oxidized substances. Oxygen thus brings forth activity, but it doesn't, however, impose its own nature on the other substance.

In teaching, it is good to discuss the content of the curriculum in such a way that relations with the nature of the human being are revealed. Here, we have reached a point where it is possible to make comparisons between the occurrences in the world of substance or matter, and human morals. If a human is capable of behaving amongst his fellows as oxygen does amongst the substances, then they would act as a great benefactor. Real selfless beneficence is characterized by helping the other in such a way, that he is able to develop his own life and personality.

The fact that oxygen follows the structure of the universe in its activities, is seen by observing the process of acid and alkaline formation and other processes.[6]

We often see that oxygen compounds generally tend to form large complexes. That is self evident, because activation through oxygen means nothing else than the manifestation of a tendency to combine with other substances.

This awakening of activity, mobility, and interaction rests partly on the principle of reaction, as such. The combining of two substances already self-evidently means the conquest of isolation and, therefore, generally implies a greater mobility and activity. We can say of oxygen that it far surpasses other substances in exhibiting this general phenomenon. What is more or less a general rule becomes here a special property of a specific substance.

Oxygen has the tendency to combine with many more substances than any other substance, with the exception of the halogens.

A major property of oxygen is promoting chemical combinations.

Fire as an Image of the Sun

We can demonstrate the relationship between fire and sun with a simple chemical experiment. For example, set up a jet of very hot flame by adding pure oxygen through a narrow glass tube to a Bunsen flame. As soon as it is applied to a piece of wood, and the wood glows and burns, we remove the flame and continue to apply the oxygen jet to the burning area. We then get a dazzling shining disk with a garland of shooting sparks. The image becomes more distinct or apparent if we use a piece of charcoal.

We have here a very striking example which illustrates the general rule: every substance in the elemental circle has properties, which makes it possible to form an image of that area from which it stems.

We now learn to see fire as a sort of image of the sun, however often it may be modified.

We may even say that every kind of heat radiation, which arises through the effect of oxygen, more or less imitates the sun's radiation, even if it is hardly apparent.

After we have seen and understood these things, it isn't very hard any more to recognize in all properties of oxygen, metamorphoses of the activity of the sun.

In many cases even the history of science seems to be determined by such a relationship. The discovery of oxygen is a very remarkable case in point. Priestley liberated oxygen with the help of sunlight from mercuric oxide.[7]

Hydrogen

Hydrogen in the Atmosphere

In comparison with other substances, hydrogen is the least visible, tangible, or weigh-able, and it is very hard to manipulate. That is due partly to its properties, partly to its scarcity.

Hydrogen is found in nature as a gas only in very low concentrations. Near the surface of the earth it is completely masked by other atmospheric gasses, which are themselves hardly perceivable. Hydrogen only accounts for about 0.01% of the volume of air.

From time to time we find hydrogen gas more concentrated, e.g., in natural gas deposits. It also is liberated during certain processes of decay. It never remains idle at one point, as carbon dioxide does, for example, until it slowly mixes with the surrounding air; rather, hydrogen disperses quickly in all directions and also, as long as it hasn't intermixed with the other gasses, rises quickly. Thus, it retreats as quickly as possible from a state where, to a certain extent, it is just perceptible to us, to hide itself amongst other substances.

This rising of hydrogen can be demonstrated clearly by filling soap bubbles with hydrogen.

In all events hydrogen exhibits the urge to reach the periphery.

Nearly all properties that make a gas observable are barely evident in hydrogen. If we collect hydrogen and compress it into a closed space, then it will meet our action with the same counter pressure as any other gas. However, if we could move our hand in that space, then we would experience even less resistance than in air.

If carbon in the form of a diamond gives us the most resistance of all substances, then hydrogen is the one that retreats more than any other substance; it continuously slips away and makes itself unnoticeable. Hydrogen is a substance that gives the impression of continuously fleeing away as if it is hiding.

Hydrogen as a Gas

Hydrogen is nearly an ideal gas. It surpasses all other low density and mobile substances in density and mobility. Every substance that changes from solid to liquid and from liquid to gas, is on its way to achieve something hydrogen exhibits in extreme. During this journey a substance frees itself from bondage to the earth, from heaviness and rigidity.

What a strange thing such a gas is! If we could let it go in an unbounded, empty space, then it would expand very quickly and thereby, of its own accord, it would rarefy itself. It occupies an ever-growing volume, but at the same time its presence at any one point diminishes. Of all substances, hydrogen is "diluted" most strongly and in addition expands most quickly.

Because the gasses near the surface of the earth are not surrounded by a vacuum, they are certainly very hindered in their movement in all directions; nevertheless, a gas isn't impeded by other gases. No gas will rest until it has uniformly filled the available space. Hydrogen will be impeded least of all—it is quickest at diffusing through other gasses.

In industry it is often necessary to manipulate hydrogen at high temperatures and pressures, for example, in the production of ammonia. That always causes great difficulties, because it is very hard to contain hydrogen. Even a metal as dense as iron becomes permeable for hydrogen! Hydrogen breaks through every kind of form and boundary more than other substances.

Hydrogen and Heat

Hydrogen behaves in every way as if it were primarily determined by heat. The "lightness" (low density), the mobility, and ability to expand quickly, are properties which are also exhibited by other substances, but only at much higher temperatures to the same degree. Hydrogen is a substance that behaves at low temperatures as if it had a very high temperature.

On heating, the temperature of hydrogen rises quickly. If we have equal volumes of different gasses at the same pressure and temperature, then, when equal amounts of heat are supplied, the rise in temperature of hydrogen will be largest.[8] That also demonstrates how hydrogen follows the impulse of heat, for a rise in temperature indicates a greater expansion rate, a greater speed of ascent. The specific heat of hydrogen is not particularly small. For practical purposes, the *rate* of heating is more important.

It is very hard to liquefy or solidify hydrogen by means of pressure increase or lowering of the temperature. That seems an obvious implication, since changing to a liquid state, a substance withdraws itself more and more from the influence of warmth. Thus, hydrogen shows the greatest contrast to carbon, which hardly will leave the solid state and, thus, in a manner of speaking, shows that it is only little influenced by heat.

Also, burning hydrogen predominantly evolves heat. When hydrogen is lit, we hardly observe any light phenomena, but a great amount of heat is immediately noticeable – a large amount of heat is evolved and the temperature rises quickly.[9] The hydrogen flame is a hardly visible, deep blue violet color.

Hydrogen and the Decomposition of Living Substances

We have learned to recognize the easy mobility and quick expansion as characteristic properties of hydrogen. It goes without saying that hydrogen will impart this property more or less to its compounds; then, even carbon will give up its rigidity. Marsh gas (methane CH_4) is a gas with a very low density. A substance like ammonia gas, a nitrogen-hydrogen compound (NH_3), is less dense and, therefore, more mobile than nitrogen.

On the whole, compounds of hydrogen with other chemical elements are gaseous, even if the other elements are normally solid.

It is not surprising that nature uses these properties of hydrogen also in the realm of life. As long as an organism is growing, and, thus, new substance is being organized into certain forms, it is organized around carbon as a center. However, as soon as an organism dies, and the substances must be returned to the surroundings, then hydrogen asserts itself and carries everything towards the periphery. A whole range of substances disappear as hydrides. Hydrogen sulfide and ammonia can often be smelled where something putrefies. In bogs and mud ditches we often notice bubbles of marsh gas, a hydrocarbon. Hydrogen phosphide (phosphorus hydride) can also occur.

Hydrogen thus plays an important part in the dissolution of living organisms, in the process of decay.

Hydrogen and Water

One of the most surprising phenomena is the birth of water from the dark glow of the hydrogen flame. We should always try to really immerse ourselves in such a process.

Initially we have the most shapeless, ungraspable, unweigh-able substance that we can imagine, something which we might call materialized "flight to the periphery." This something ignites with an intense heat, and a vapor evolves which, as long as it is hot, expands as a gas, but if we cool the surroundings, condenses to a mist or even to droplets.

Just as weakly as hydrogen, water has few distinct qualities. It hardly has a color of its own, little smell or taste, and little chemical activity. While hydrogen, however, brushes past everything without being influenced, water will readily dissolve all manner of substances, by which it can take up all sorts of qualities.

It appears as if nature has given water the task to grant other substances the opportunity to unfold their own nature to the fullest extent. Water is like a kind of crystalline fairy castle, in which many objects that had been under a spell of rigidity outside, once inside they are freed from their spell and can achieve interaction.

With the change from hydrogen to water, polar opposite properties emerge.

The contrast between hydrogen and water becomes even greater when we look at their reaction towards heat. While hydrogen reacts most quickly to heat of all substances, and, thus, a small warming gives rise to a strong reaction and temperature rise, water is hardest to heat up. For water, we could speak of an unusual inertness or even of resistance concerning heat: we must supply an extraordinary amount of heat to make ice melt, to raise the temperature of water, to force water into a gas.

We also need an exceptional amount of heat to decompose water into hydrogen and oxygen.

In summary we can say: more than any other substance water has the ability to use heat. But precisely this property is very important for the earth, particularly for regulating and harmonizing the climate. All the heat water absorbs, without undue change, it can also again give off somewhere else and under different circumstances. Thus, it is also the best exchanger and transporter of heat.

By comparison, hydrogen gas is most inefficient in transporting and absorbing heat. Just as its temperature will rise quickly after a absorbing a small amount of heat, so only a small amount of heat will be released, even if the temperature falls quickly.

The substance which is most affected by heat, is the least capable of absorbing and transporting heat. More than other substances water, which is least affected by heat, is capable of absorbing and transporting it. The fact that blood can maintain its constant temperature rests for a great part on this remarkable behavior of water. In us water is the calm transporter of heat.

In this marvelous fall into its opposite, which occurs during combustion of hydrogen, we must acknowledge the influence of oxygen. Henderson[10] has pointed out that in the production of water we have quite an exceptional phenomena. As a rule, hydrogen compounds of an element are more volatile than the element itself. However, water is much less volatile than oxygen is. Henderson has also already indicated that because of this anomalous behavior, water can form the basis and environment for life as we know it. We may add that its importance for chemistry is related to this, also.

In water, we can clearly recognize both the respective tendencies connected to hydrogen and oxygen. As soon as water rises and vanishes, we can see the influence of hydrogen. When the vapor condenses and forms clouds and seeks the earth anew, then we can recognize the influence of oxygen, which maintains the connection with the earth.

Water is the most suitable medium for chemical reactions. This also is due to the two main properties. In the first place, water is the best solvent there is. As a consequence it can bring the most diverse substances into intimate contact and interpenetration. In the second place it has the unique gift of enhancing chemical reactions. In the first property, we may recognize the ability of hydrogen to bring substances beyond their boundaries and make them susceptible to change. In the second, we may recognize the ability of oxygen to bring forth activity and interaction.

This task of water to open the way for chemical reactions to occur, and it is so specific that it seems impossible for a chemical reaction to take place without at least a trace of water being present.[11] Even a mixture of hydrogen and oxygen will not explode if the mixture is absolutely dry. It will burn, but with the speed of a slow fuse. It can burn steadily due to the water being formed.

In the language of the alchemists we would say: In nature we find that the mercurial principle is especially represented by water. It mediates between extremes which it brings into contact and interaction. Everywhere its influence is harmonizing, or—to use an expression from human life—equalizing/balancing, and beneficial.

Previously we saw that its beneficial influence is due to the fact that water itself is a product of a harmonizing process. In water, we see nature solving one of its problems.

Perhaps a characteristic trait of the earth is that life is only possible on the basis of such a substance. Life on earth is only possible with harmonizing, equilibrium-creating forces in the background.

Carbon

Carbon occurs in nature in three different forms: diamond, graphite, and amorphous carbon. Diamond, the most noble form, is rare. Amorphous carbon includes all non-crystalline types of carbon. Although they differ very much, they have a common factor: form and properties are strongly determined by the manner of their creation. If we include substances like coal, which are in the process of becoming carbon, then we find considerable amounts of amorphous carbon.

Diamond and Graphite

A comparison between the properties of diamond and graphite gives us the opportunity to mention a certain method in teaching. A lot can be told about diamond and graphite, if we consider their usages and manufacture. However, if we particularly want to review the properties of the substances themselves, then we quickly come to an end, except if we approach it via a certain path. This path is to describe the interactions of the substance with the various forces of nature and with other substances. So, in each case we will emphasize a certain property in a more or less abstract way, and immediately illustrate it in the manner indicated.

We'll begin with diamond, the most pure form of carbon. Diamond occurs as a cube or an octahedron, i.e. in the most compact, earthly crystal forms imaginable. The unpolished diamond is often somewhat cloudy on its surface and not very remarkable. Diamond surpasses all other substances in hardness. This means that there is no object to be found on earth which could make even the smallest scratch on its surface. However, with a splinter of diamond one can scratch on any surface, even the hardest. Diamonds thus have one of the characteristic properties of the Earth element, namely the ability the retain its specific form, to a very high degree.

When the unrefined diamond is cut, then a substance appears which like many crystalline substances is very transparent and has the greatest refractive power. On the one hand, it leaves the field clear for light, and on the other, it influences light more strongly than any other substance. If one would make a diamond prism, then light would be extremely refracted, and it would appear with wide fringes of color.

If we look through any crystal, under certain circumstances the rear face will act as a mirror—it appears opaque and presents an image of what stands in front of it. The phenomena of total reflection is very strong with diamond, in accordance with its strong refractory power. If we shine light onto a cut diamond, a play of light starts which is more brilliant than with any other substance. Because of strong internal reflection, light is more or less "caught" in the diamond. If it then does shine forth, it does so in strong warm colors. It appears as if the impulse which rules the Earth element — namely to be isolated and to contract to its specific form—is being extended to the light.

Graphite is much more abundant than diamond. But despite this fact and that its much less noble appearance makes it a much more common substance, it still has very remarkable properties.

Its color is gray-black, which means that there is no play with light possible; to the contrary, it destroys light. More or less in contradiction to this is that graphite glistens and reflects quite strongly, which means that it throws back and reflects light, just as a metal does.

Graphite is one of the softest of mineral substances. It will even impart its black color to soft paper, although if the graphite of our pencils didn't contain some clay, they would be unsuitable for their task. Not only because of its softness, but also because of its peculiar crystal structure, graphite can be used as a lubricant—a *solid* lubricant! It is used in pianos where wooden parts of the keyboard have to slide over each other with minimal resistance and wear. It is also used when graphite is mixed with cylinder oil for engines, and also for locks. From this it becomes clear that with the softness of graphite we are dealing with a case of extraordinary adaptability. If we rub a rough surface with a graphite crystal, then all the openings will be filled, all the roughness will vanish, until the surface is as smooth as the graphite crystal itself.

If we summarize these facts in thinking and draw a comparison between the properties of diamond and those of graphite, then something surprising emerges. It seems as if great care has been taken to create two substances, which, to the smallest detail, are opposites. We face here one of those great moments of equilibrium which are typical for the order and harmony in nature. More so than other substances, carbon expresses itself in extremes; however, in this case they are exceptionally well-attuned—and that indicates the central role of carbon.

It makes a difference if we can use a crystal to scratch every other substance, or if something of one's own substance is given off to every slightly hard surface. In the first case, form retains its substance with unwavering force and thus opposes every resistance; in the second case the smallest influence is sufficient to cause the substance to lose its form. And yet graphite exists in a specific crystal form, although different from that of diamond.

A further contrast: diamond powder, the most powerful abrasive, is best suited to grind and polish a surface, and a graphite film, which forms a cohesive, tangibly slippery mass eliminates nearly all roughness.

The use of diamond in drilling into hard rocks where everything must yield to its hardness is interesting. And yet diamond is so brittle that we can crush and pulverize it in a steel mortar; even this hardest substance must give way to a superior force. However, it does not yield; it is more like retreating in ever smaller pieces, which, however, retain the same facial angles. In every respect, diamond emphasizes separation into a specific form.

We could say that carbon carries through to an extreme, a certain aspect of the Earth element. Graphite, however, exhibits some contradiction in its properties. It is a solid, but it doesn't insist on separation. It adapts easily to its environment and unites quickly with other substances. It is self evident that diamond is a bad conductor: a solid that is opaque doesn't conduct electricity. It's just as clear that graphite, being a good conductor, would be opaque. In this it resembles the metals - where electricity can freely act, light is excluded.

Through the remarkable combinations of its properties, graphite is the ideal substance to be used in industry when a non-conducting object has to be covered with a coating of metal. For example, one first rubs a gypsum sculpture with graphite, thus giving it a continuous, conducting skin. Now this figure can be covered electrolytically by a metal layer. This method is called "galvano-modelling."

How Carbon Withdraws Itself from Life and Chemistry

The process of life, especially as it manifests itself in plants, has something very delicate, a finely mobile and dynamic equilibrium. We can experience, as directly related, the way in which oxygen is released by the plant as a beneficial flow under the influence of the sun.

How different when carbon is released from a lifeless being, by scorching. We then get a substance which remains unchanged under the influence of air and water. While oxygen acts as an awakener of life, carbon, because of its rigidity and isolation, may be seen as the natural symbol for death. And yet, carbon forms the center of nearly all substances which are important to life. Nearly all these substances are carbon compounds.

It is even possible that by heating parts of living organisms in exclusion from air, we get crumbly pieces of carbon which, to a large extent, have retained their original shape. This sort of carbon which doesn't have an apparent crystal structure but gets its form from influences which are not related to its own laws, is called amorphous carbon.[12]

The substances which are connected with life processes are often found to be fluid or are even gaseous. Others are solid, but condensed from the fluid state. Most of these substances are very susceptible to influences and very changeable.

Carbon as element doesn't have any of these properties. It is very hard to induce any change in carbon through heating, through dissolving, or through the action of chemicals. (That also applies to diamond and graphite.)

If we heat a piece of coal or charcoal in the absence of air, it remains dazzlingly glowing and solid, while most other substances would have melted or even volatilized. Only above 3500°C will it change to a vapor, hardly going through the liquid state. That is about the most tenacious effort of trying to keep one's own form and boundaries we know of. We know of no solvent in which carbon can be dissolved at normal temperatures. However, there are some metals which when liquid, will dissolve carbon and thus activate it. Of great technical importance is the way it dissolves in the iron flowing from a blast furnace.

Under normal temperatures fluorine is the only substance that will attack pure carbon. (And that can only occur in the laboratory.)

All other chemical influences are warded off. That is why we can use pure ("refractory") carbon when electrolyzing chlorides. The liberated chlorine, which would attack every metal, doesn't damage the carbon electrode. Also, we can protect wooden piles against decay and the effects of the soil by charring them in a fire.

The only way to activate carbon is by combustion. Above a specific temperature, which varies for the various types of coal, will the carbon react with oxygen. The result is carbon dioxide (carbonic gas), a substance which in nearly all respects has opposite properties from carbon.

The typical properties of carbon indicate that as an element it has an almost exclusive relationship to Earth. Even in compounds carbon is still filled with the impulse to go back to an earth-condition. The many natural processes, such as carbonization or a sooty flame, which tend to liberate carbon black, indicate this. Precisely this substance which is involved in one of the most remarkable (matter) cycles in nature, has the tendency to isolate itself and to withdraw from this cycle.

We should, nevertheless, remind ourselves of the vast distance there is between the living substance and pure carbon. All intermediate states, such as charcoal, are much more receptive to chemical action than is carbon.

Phenomena with Combustion of Carbon— "Right Concepts" in Science and Technology

Considering the almost total chemical inactivity of carbon with respect to most other substances, it is surprising that its ignition temperature is relatively low. While for diamond this temperature is 850°C, graphite will ignite at about 700°C, and amorphous carbon often at temperatures of around 200°C. Hold a piece of carbon, for example, in a gas flame. The black thing then glows orange-colored, even when the flame is taken away. We observe how the black mass is being consumed, while a very light blue-gray ash remains. Investigation reveals that oxygen from the air is being used and carbon dioxide produced. We thus observe carbon disappearing and reacting with oxygen in a glow, from which heat and light arise, but also invisible carbon dioxide is liberated, which light and heat have abandoned.

In our experiments, we should always take note whether we are dealing with a whole or with only a fragment. If we want to give a demonstration of the typical combustion of carbon, then we should ignite a pile of wood from below and from the side. If we then forcefully blow into the fire, then a very hot space will arise in the middle, with its walls glowing a yellow-red, while the pieces at the outside remain dark. Carbon thus gives us a very characteristic picture of combustion, which other substances never show – but a piece of coal by itself only shows a fragment of the total picture.

It is, of course, possible to describe the combustion of carbon in the usual way: under influence of high temperature, carbon combines with oxygen to form carbon dioxide, and during this process, a lot of heat is created.

$$\text{carbon} + \text{oxygen} \rightarrow \text{carbon dioxide (+ heat)}$$

$$C + O_2 \rightarrow CO_2 \quad (+ \text{A kilo-calories})$$

By proceeding in this way, we are saying something factual, but we are neglecting an important part of reality. We completely blur the difference with regard to the combustion of other substances. However, if we really attend impartially to what our senses give, then we find: coal is consumed by a fiery incandescent glow, so that a coal fire continuously increases the inner space. Other substances, like phosphorus or sulfur, give the impression as if the substance is "going up" in the fire, while here the fire is attacking and *consuming* substance. Take these indications as guidelines for careful observation of the phenomena yourself, and it will then become apparent what is meant. Even by burning wood or paper, one can observe what is necessary.

It may seem unnecessary to explain this with so much emphasis. However, we must always be aware of the schooling of the human soul, and this includes the awakening of "right concepts." Such concepts may become windows through which we may see a deeper layer of reality, while incorrect or abstract concepts may be compared to a thick wall, which we ourselves throw up in front of this deeper reality. Moreover, only right concepts may be taken into sleep to have a productive and constructive influence. These are related to the finer growth forces and thus open the possibility for us to influence and enhance them.

It is nearly self evident that our technical equipment, as far as it functions properly, should reflect these right concepts. In every oven the hottest place is in the center, yet often on the outside the coals may still be black. They always burn around small air pockets or fall into the glowing center to be burned. It is quite different to burn (pieces of) phosphorus or sulfur in such an oven. The study of technology can be an important aid to correct abstract scientific concepts. If we first stimulate the glowing mass of coal by blowing and then stop, we will see small pale flames shoot out. The reason is that carbon reacts with oxygen in two stages. If the air supply is insufficient, then carbon monoxide, which can subsequently react to carbon dioxide or carbonic acid gas, is formed. By fanning the fire, the heat becomes very intense. If we then stop fanning, then the amount of oxygen is insufficient and a shortage is created. At the glowing surface carbon monoxide is formed, which flows out and catches flame as soon as there is sufficient air once again.

As a rule, flame phenomena occur because heat liberates combustible gases or vapors from the fuel. Because carbon isn't volatile at the combustion temperature, it cannot actually develop flames. However, it becomes possible through the generation of carbon monoxide gas. Through a very striking example, we can see again how processes in nature are not governed by various laws which can be expressed in abstract terms, but at the very least also are guided by something we could call "laws of appearance." Carbon is a fuel and, thus, should be able to show flames. And this is accomplished here, even though carbon itself doesn't become volatile and cannot liberate gasses.

On Carbon Dioxide

When carbon changes to carbon dioxide, this results in a complete reversal of nearly all of its properties. The rigidity becomes great mobility and leanness, passivity becomes many-sided activity, and the gesture of confining oneself in one's own boundaries becomes one of participating in the formation of many substances.

But even then, a certain particular affinity to the earth remains. That becomes apparent by comparing it to other gasses in the air. Carbon dioxide is the heaviest gas among them. If we release pure carbon dioxide in air, it will immediately sink down.

We can easily demonstrate this by setting up a hydrometer cylinder into which we blow smoke through a long glass tube. Now as we "pour" some carbon dioxide into the opening of the cylinder, we observe a vortex moving downwards. Then, before too much smoke has dispersed, we tilt the cylinder towards a horizontal position, and then we observe the smoke flow out and sink down as if it were water—the smoke has mixed with the CO_2 gas and now reveals the path it takes in sinking.

We can also do this with an empty smoke-filled aquarium, and subsequently let carbon dioxide spread itself along the bottom. If we now move the aquarium to and fro we see the smoke come into wave motion, as if it were floating on a liquid. It is, of course, also possible to do the classic experiment with the candle, which is extinguished as it sinks into the carbon dioxide layer.

It is very impressive to fill soap bubbles with carbon dioxide from a tank; they fall to the ground almost like stones.

All these experiments can also be done with hydrogen (except immersing the candle). But, everything then has to be done upside down, and the whole sequence of events is then also reversed. As we will see, this is due to a real connection between carbon dioxide and hydrogen.

Also, in the distribution of carbon dioxide in the atmosphere, we find indications, which lead to the conclusion that there is an affinity to earth. The atmospheric content of carbon dioxide is very small, about 0.04% (volume %). At a certain distance above the soil this content is practically constant; but, it rises nearer to the living soil. It is quite common for carbon dioxide to be liberated from the soil, or to be released with so-called "mineral water."

Not only does carbon dioxide have a special relationship to earth, but also to water. It is much more soluble in water than the other atmospheric gasses. That specially applies to water containing lime and, above all, also to sea water. If for any reason the carbon dioxide content of the atmosphere would increase, then the sea water would immediately take up the excess (change of carbonates to bicarbonates). If there were to be a shortage of carbon dioxide, then sea water could liberate great quantities.

However, it appears that the sea water nowadays cannot cope with the excess of carbon dioxide produced by industry and transport.

Such a relation between carbon dioxide and water is not only of importance for nature, but also for the food industry. If carbon dioxide stands above water, then under increased pressure more and more carbon dioxide will dissolve in it. If we then decrease the pressure, it will effervesce (form bubbles), as we can observe with mineral water and other carbonated drinks. This effervescence of carbon dioxide (which, of course, can be seen to a greater or lesser extent with other gasses) but which is very pronounced here, I consider a particularly important phenomena which should be very carefully studied. In a way it is the opposite of drop forming. These two (drops and bubbles) are some of the most remarkable "mercurial" phenomena. We can observe this bubbling repeatedly and without much effort by adding a trace of acid to soda water solution.

The dissolved carbon dioxide gives the water the well-known prickling, stimulating taste. The solution is slightly acidic, which indicates the formation of an unstable acid, carbonic acid.

Not only does carbon dioxide show a clear affinity to the earth and water element, but, in contrast to the other atmospheric gasses, it is also easily liquefied by pressure increase. It is traded as a liquid.

By holding a cylinder filled with liquid carbon dioxide with its nozzle pointing downwards, so that after opening the valve liquid flows out, then we can change it to solid carbon dioxide (carbon dioxide "snow") by catching it in a woolen cloth or bag. Because of the swift expansion, the temperature plunges so quickly that the substance partly freezes.

If we consider the role of carbon dioxide throughout the whole earth, then we must conclude that this substance contributes more than any other substance—with the exception of oxygen and water – to the massive conversion of matter. Carbon dioxide is a weak acid, but its effects are much more extensive and drastic then those of the strong acids.

In the first place carbon dioxide plays a large role in life's processes. It is exhaled by all living organisms and is one of the most important ingredients of photosynthesis.

Its effect on minerals is more of a *chemical* sort. It promotes the weathering of igneous rock, because it combines with the alkaline components of the rock and forms soluble salts which are washed away. The remains of the rock begin to disintegrate.[13]

In limestone, which predominantly consists of calcium carbonate, massive amounts of carbon dioxide are fixed in the earth. We should really be very surprised that this rock is also attacked by carbon dioxide-containing water; it is dissolved as bicarbonate and washed away by the flowing water.

Both in the case of the weathering of the igneous rock as with dissolving of limestone, carbon dioxide carries out a tremendous attack on the structure of the crust of the earth, as it has been preserved from the past until now. The new compounds which are formed flow predominantly into the water sphere.

In the plants, carbon dioxide aids the synthesis of new forms, whereby there always is a transition from water to earth.

The importance of Carbon Dioxide and Carbon for the Life-processes in the Plant

The most important processes in which carbon takes part occur in living organisms.

During photosynthesis carbon dioxide follows the typical path of carbon from air to earth, from light to darkness.

The plant needs carbon. In a certain sense the plant is even organized and extremely many-sided, differentiated carbon. The plant owes its sharply defined form and rigidity, its ability to maintain highly complex structures, to a large extent to the propensity of carbon to become compact and rigid.

The plant absorbs carbon dioxide from the atmosphere as a very mobile gas. The plant draws, thereby, from a source which renews itself ever again. That should be considered a great advantage, while initially it could be seen as a disadvantage that carbon dioxide occurs in the atmosphere in such a small concentration. But just this circumstance allows the plant to exercise its "calling" in the most intensive manner. Indeed, it is the role of the plant to give structure and form to amorphous substances. The more rarefied and mobile the raw material is, and the longer the passage from initial amorphous to complex end-state, the more impressive is the efficiency of the life-processes.

We could describe carbon dioxide as carbon awakened and brought close to life by oxygen. Photosynthesis begins with a reduction. While light and water are being assimilated, oxygen is being liberated and with this, activity, and mobility. It is really surprising that starch is the first substance which can be clearly distinguished. After carbon dioxide has penetrated into the plant through the stomata at the bottom of the leaves and has dissolved in the liquid, then it rather quickly is changed into starch. Carbon dioxide is changed into a solid, granular state, just about the opposite to its state as a gas.[14]

For starch to be active in the life-processes of the plant, it has to be converted to glucose; this happens all the time, and starch is used as a reserve.

The formula for glucose, which is the sugar concerned, is $C_6H_{12}O_6$. We could also write this as: $(CH_2O)_6$. Therefore, we can consider sugar as a substance in which carbon in combined with water, the substance which offers the real working-environment for life's processes. Through the reduction process, sugar acquires its combustibility; through its water content, it is able to participate in a highly differentiated way in many vital processes. The aggressiveness of carbon dioxide has disappeared and has given way to an open passivity, to responsiveness to vital influences. As far as this goes, sugar in the plant surpasses all other forms in which carbon appears.

There still exists a certain resemblance of sugar to carbon dioxide in so far as it is present in a mobile, very dilute state.

Of great significance is the plant's ability to condense sugar to starch and cellulose, without markedly changing the composition. A certain amount of water, however, is released, and that does, of course, contribute to the solidification.

The construction of the plant form is largely the result of the thickening of sugar to starch and wood. At the centers of growth, sugar leaves the liquid state and changes into more or less solid, insoluble substances. The sugar is ejected from the domain where vital forces are fully active to the realm of the mineral forces.

The composition of cellulose is still very similar to that of sugar. In pure cellulose, as in cotton fiber and flax fiber, it possible to see that light contributed much to its creation. However, some of the main characteristics of sugar, its solubility and accessibility to chemical influences, have changed to their opposites. Cellulose is a very stable substance. We could even call it a noble substance, in the same sense as we speak of noble metals. In this stability the specific properties of carbon clearly come to the fore.

If cellulose is condensed further so that wood is formed, then softness, elasticity, and permeability for water are lost. The stiffening is taken further.

Whereas it is characteristic for sugar to be very amenable to the many varied formative tendencies, it is typical to find in the various cellulose structures permanent imprints of this malleability. We see here a major property of carbon: the ability to be impressed by the forces of life and to retain these forms.

In a growing plant, initially only a germ, the seed unfolds its form more and more into space. This process is diametrically opposed to the contracting forces of the earth. However, the carbon which is incorporated into this form, initially from a very rarefied, finely dispersed condition in space, becomes condensed and hardened. If we contemplate these remarkable phenomena, then we discover the surprising fact that water in the plant primarily relates to growth, while carbon moves in an opposite direction.

This points to something essential. Rudolf Steiner repeatedly points out how those forces which make a substance suitable for life, which have a molding influence, are so very difficult to grasp for present day thinking; this is because they act from the periphery of the earth and are in every respect the opposite of the typical forces one encounters in the inorganic world—these act from a point-center. The resulting growth of the plant is the consequence of a sort of battle between the cosmic-peripheral forces, and the terrestrial-central forces.

The water in the plant is the basis for the action of the vital forces. As a consequence it even moves in the same sense in the spatial direction. To a certain degree that part of the water which in the process becomes built into the plant substance is being held by the forces of earth. Carbon is captured by the forces of earth but in such a manner, that the forms and structures it takes on are determined by necessities of life.

Water flows away from the realm of darkness and gravity towards light, from compactness to levity. Carbon sinks from the realm of light into that of darkness, from clear lightness into dense darkness, and also from the realm of flowing forces of life into the sphere where the mineral forces predominate. We could also say: Initially carbon takes part in processes that proceed in time; then it coagulates into forms in space. But the resultant form is, as it were, a projection of the flow of time into space. With plants, it is always so, that what has come about in a sequence of time will be preserved in a static form in space.

Thus, one of the main properties of carbon is to form the basis for occurrences which take their course in time to thicken into spatial structures.

This understanding should become one of the points of view through which we can come to a more living concept of structural formulas in carbon chemistry. We can view these partly as a spatial depiction of processes which have occurred after each other in time.[15]

Everything of a living being we can see and can touch is really already an end-product. With our senses we don't observe the process of life, but only its imprint on matter. It is above all carbon that helps living creatures make themselves visible to our sense organs. Or, put differently, the process of life, which takes place in time, becomes a matter-filled, spatial shape.

If we really understand the mission of carbon, then the properties it has as an element become clear. We can now understand how carbon as diamond maintains itself so well; that carbon has the tendency to fall out of mobile processes and congeal in the amorphous state; that it is very hard to activate it as an element and that it stubbornly retains its original state.

Carbon between Sun and Earth

Elemental carbon is mostly black. A typical carbon compound, such as sugar, is colorless or white when reduced to a powder. When heated, the color changes to brown and then to black. To a large extent the processes in carbon chemistry express themselves in such color transitions between black and white. A white or colorless, transparent substance exceeds the other variously colored substances in its ability to play with light; a black substance absorbs light.

With carbon as diamond we can bring this play with light to its zenith. If we really want to eliminate the influences of light, then we have to use carbon as soot; this is the most black substance known. The most wide-spread use of this property is in printer's or India ink.

Every substance nullifies light to a certain extent. It is very hard to notice that even air dims the light passing through it somewhat. Water and other liquids, specially when they are colored, show this phenomenon much more clearly. With transparent solid substances this is even more noticeable.

When solids have the tendency to exclude the light from the space they occupy, then we call such substances opaque. However it happens, light always loses intensity and effectiveness when it comes into contact with matter. We call this absorption of light. However, we should really speak of an activity of darkness, which emanates from matter, because when light is destroyed or is kept away, darkness comes about.

Carbon can be the bearer of exceptionally strong darkening powers.

Also, in this respect carbon has a property which is characteristic for all substances, only much stronger than any other substance. Herein carbon expresses its special relationship to the earth element.

We describe the phenomenon of light absorption correctly only if we mention that to the same degree that light is obliterated, heat emerges. At least that is so in general; other effects also come about but to a lesser degree.

In its black form carbon has the special capacity to obliterate light and create warmth.

When various substances are heated, then the properties we've been discussing turn into their opposites (providing the high temperature doesn't bring about other drastic changes). For every substance there is a certain temperature at which it will start to glow. In other words under the influence of warmth, light phenomena become visible in the dark. In general the rule applies here that substances which when cold only absorb a little light, when heated will only produce a little light. In contrast black carbon begins to produce light at a lower temperature than other substances, and radiates more light at the same temperature.

At higher temperatures carbon has the special ability to produce light at the expense of heat. At lower temperatures it has the ability to generate heat at the expense of light.

We can easily demonstrate such phenomena. Take a piece of charcoal (for example, a carbon rod from an arc-lamp) and draw a few lines on it with chalk. If we subsequently heat the carbon in the dark, then we will observe that the black surface starts to glow, whereas the stripes remain dark. If we then illuminate the rod with a powerful light source when it doesn't glow strongly any more, then the stripes are suddenly white against a dark background. We can do the opposite experiment by drawing lines with some graphite on a piece of white porcelain. In this way it should even be possible to write with flaming letters. That will probably be achieved successfully by using a piece of rough-surfaced silver and graphite.

It requires a special skill to be able to describe and characterize laws and phenomena of nature. The understanding we arouse in the end depends to what degree we have developed this skill. It can be very fruitful to give more than one description if it isn't possible to characterize the phenomena with one.

In these paragraphs we have something of great importance before us: certain phenomena appear in the realm of the visible, others disappear. On a higher level we find the same with birth and death. We can thus use expressions from this sphere to characterize those from the other. We then get the following:

In the black carbon surface, light dies in all its expressions, while warmth (heat) is produced. In the glowing carbon surface light is created while heat dies.

In these properties carbon is the complete opposite of silver. There is probably no other substance which reflects light (and its manifestations) so totally and which takes so long to produce a glow as silver.

It is said that after hydrogen carbon has the highest heat of combustion. This is the heat in calories produced per kg of reaction product. We could, of course, also measure the heat of combustion per unit of volume or per mole and would then obtain other ratios.[16] At any rate, the amount of heat produced in combustion is very large. Therefore, it isn't quite correct to say that carbon is related to the earth element. Carbon is warmth-charged earth, quite different, for example, from quartz, which is pure earth.

If we ask ourselves what the ability of carbon to accumulate large amounts of heat relates to, then we have to look for the answer to photosynthesis. During that process radiation of the sun is absorbed, while at the same time the basis is laid for the rigidification and separation of carbon. During combustion the opposite occurs. A powerful glow is created, through which carbon becomes very mobile and vanishes into space. What initially had been fixed in matter as sunlight, becomes free again. We are not only dealing with accidental energy changes, but we must see carbon as a substance which has the special ability to connect sun-impulses to the earth. We shouldn't be surprised, therefore, that with this substance it is easy to create the image of the sun.[17]

This is also the reason why carbon not only plays a main role in heating technology, but also forms the basis for the illumination industry. With fires, torches, candles, gas flames, etc., light is mainly the result of glowing carbon. The first electric incandescent light bulbs used threads of carbon. The very powerful light and high temperature of an arc lamp are achieved with carbon rods.

In the whole area of heating and light technology one is creating artificial suns, as it were. By lighting a lamp darkness is driven away as is the case when the sun rises.

When contemplating such facts, it can become particularly evident that by burning coal, the blackest substance changes to light. The blackness of coal is conquered by light, just as the sun overcomes the night. We may really take the combustion of carbon as an exact image of a sunrise. If we bring carbon to glow in any other way, then that should be characterized in the same manner. There light also comes about by conquering darkness.

In these phenomena—in this shining out of the darkness—the nature of carbon is already clearly expressed. However, this is even more so in the case of the photosynthesis. As we have previously seen the nature of oxygen is just as apparent as that of carbon during combustion, while photosynthesis and everything related primarily depends on the task of carbon.

During the life of the green plant, the sun's radiation is strongly bound to matter. Light disappears without heat being created in any large amounts, and substances come about which in their final state would be black.

Just as during combustion light emerges from the darkness, so here light is led into darkness and becomes fixed. We may, therefore, view photosynthesis and what arises from it as an image of the sunset with respect to the reactions between substances, as we saw the combustion of coal as an image of sunrise. During both synthesis and during sunset, light disappears into matter.

In the devouring of light by the blackness of carbon upon which it shines, even in the arising of soot from the fire, we have images of the victory of darkness over light, as in a sunset.

More than any other substance, carbon offers sunlight the possibility to be fixed with the solid earth and to disengage itself. It presents us in its most important processes a picture of sunset and sunrise or, in other words, the entry of the sun into the earth and the emergence of the sun from the earth.

Black coal is a picture of the earth in its night-state, when it has hidden the sun.

Carbon dioxide, which is in the atmosphere and thus is being illuminated, may be compared to the earth by day, as it is the opposite to black coal.

We have got to know carbon as a typical earth substance. And precisely this substance carries an imprint of the whole interaction between sun and earth with it. That is one of the signs that signify that in the earth there lives a hidden relationship with the sun.

If it is possible to dig up an uncut diamond from the darkness of the earth and to process it in such a manner that it displays its matchless play with sunlight, then that also indicates this. This phenomenon also signifies a hidden longing of the earth to become one with the sun. At this point we must refer to a secret which forcibly makes itself known and yet we must leave it, without solving its secret. For it is such an astonishing phenomenon that the state of a substance which can be seen as the pinnacle of all earthly existence (diamond) manifests itself in an element and not in a compound.

I do not think I am wrong in saying that in nature there is only one other element that is found in a transparent, crystalline form, and that is sulfur. Sulfur crystals are very beautiful, only one can not speak of a culmination. Anyhow, they only underline the uniqueness of the occurrence of diamond.

More on Diamond and Graphite

In diamond and graphite, we have before us the crystal forms and states of substance carbon will take on when no outer influences are active.

Do we find in these states a mirror image of the tasks carbon has to fulfill in nature?

In the first place we find something of the light-dark, of the white-black in which the chemistry of carbon moves. Diamond opens itself to light and plays with it in an unparalleled way. Graphite is very opaque and black; furthermore, it is quite a good reflector. It offers light a very limited domain. We have here the remarkable phenomenon that one and the same chemical substance manifests itself in a state that resembles the metallic state and as a transparent crystal. It has already been mentioned that these are fundamentally opposites states.[18]

This again points to the fact that carbon occupies a central place in nature.

We have seen that in the plant, carbon initially takes on a very pliable, influenceable state and then changes into a state which retains and fixes the impressed influences.

Carbon distinguishes itself, on the one hand, by being very pliable and open to influences, and, on the other, by stubbornly retaining its form accompanied by a resistance to influences. We see immediately how these properties appear in diamond and graphite in a striking manner.

Carbon — Hydrogen — Oxygen

Kinship and Contrast between Carbon and Hydrogen

When we compare the properties of carbon as element with those of hydrogen and especially their typical, natural cycles, a surprising relationship then appears. The connections between both substances show, on the one hand, a certain similarity, on the other a deep opposition.

They are combustible and both evolve when burning a lot of heat in comparison to other substances. In both cases the product of combustion is of vital importance for the processes of life and for the chemistry of the earth. We are faced here with phenomena which have to be contemplated with great seriousness: combustion accompanied by heat and light radiation and simultaneously the occurrence of a certain openness or susceptibility to influence from the realm of life and chemistry. Generally, the action of oxygen and especially combustion cause a profound enhancement of chemical activity. Rudolf Steiner talks about a contrast between chemical and heat effects.[19] This contrast is clearly expressed in the above mentioned rule. Substances like carbon and sulfur are, as it were, charged with warmth, but less susceptible to chemical influences. After ignition a lot of heat streams away, and simultaneously chemical activity emerges. We could call the powerful chemicals, like sulfuric acid or sodium hydroxide, very "burnt out." During their formation from the elements a lot of heat was evolved.

Considering the great amounts of heat that are evolved during the combustion of carbon and hydrogen, we would expect water and carbon dioxide to be chemically very active. While we do not find powerful action, we do, however, find an unusual many-sided chemical activity. But all activity proceeds very calmly.

We must now focus our attention on the following facts.

Of all elements, carbon and hydrogen exhibit the greatest heat of combustion.

Carbon dioxide and water, the products of combustion, are the most common products of combustion.

They are also the products of exhalation of all living organisms.

Both substances, after oxygen, are responsible for most of the chemical changes in nature.

Both these substances are the predominant ones used by plants for their growth.

Finally, carbon and hydrogen both show a remarkable tendency to withdraw from the great matter cycles and to persist in the elemental state. And here begins their opposition. When carbon is freed from its compounds, it moves in the direction of the center of the earth; hydrogen, however, goes towards the periphery.

Carbon can develop properties with which it specifically penetrates into the sphere of the senses. As diamond it is harder than any other substance, we could say, that it forces everything that comes in its path aside. As element hydrogen has very few properties which force themselves on the senses. No substance puts up such little resistance as does hydrogen.

Diamond retains its form-quality more than any other substance. Not only is hydrogen the least-formed substance that there is, but it also removes the formed-condition of substances it combines with. Hydrogen is, so to speak, matter imbued with a protest against everything that has form.

It is nearly impossible to get carbon out of its state of rigidity other than by combustion; carbon is rarely seen as liquid or vapor.

But hydrogen can be fixed out of its state of intense mobility only with the greatest of efforts; it is rarely seen as a liquid, let alone as a solid. Here also combustion helps and gives us a substance which can easily be solidified.

We have seen how hydrogen is related to warmth in every respect. Also, it behaves at low temperatures as other substances do at high temperatures.

Carbon is partly related to fire. However, we can also say that at high temperatures it behaves like other substances at low temperatures. This seems to indicate a relation with coldness.

Compounds of Carbon and Hydrogen

In view of the above it seems very interesting that there are countless substances which produce essentially only carbon and hydrogen upon analysis. Petroleum consists almost entirely of mixtures of such hydrocarbons. Nearly the same applies to tar products.

We want to restrict ourselves to those components of petroleum which are classified as "saturated" hydrocarbons.

With these substances the characteristic of combustibility comes to the fore in a very one-sided way. On combustion they produce a lot of heat. At the same time they are chemically so passive to chemical influences that they are called "paraffins."[20] In the light of the previous discussion this combination of concentrated heat-energy and chemical inertness shouldn't come as a surprise.

This is accompanied by a poor relationship with water, which is not only evident in its insolubility, but also in the ability to repel water.

These petroleum products are found mostly under the crust of the earth. They have followed the direction of carbon during the geological development of the earth. Usually the are subjected to immense pressure, which can even make them gush up when tapped; here again they imitate hydrogen. If, furthermore, we take into account that they form a series with gasses at the beginning, liquids in the middle, and solids like paraffin or asphalt at the end, then we notice that they form a sort of bridge between hydrogen and carbon.

Although these hydrocarbons are unsuitable for nearly all living beings, substances like fats and oils made during the process of life are not so toxic. Here we also find a strong concentration of heat, water-repelling properties, and quite a strong chemical inertness. With some effort every fat may be converted into a fatty acid and glycerin. The fatty acids, even more than the fats, are closely related to the hydrocarbons. However, they do contain some oxygen and, therefore, offer a point of attack for chemical influences and life.

Glycerin is comparable to the lighter hydrocarbons. It distinguishes itself from them by its large oxygen content. It is combustible, at the same time chemically labile, and dissolves easily in water.

Due to their aliphatic-like (hydrocarbon-like) nature, fats are of great importance in the warmth/heat metabolism (and for repelling water e.g. the feathers of the birds). Due to their remarkable composition and their oxygen content, they are susceptible to life-processes.

The essential oils, these volatile, fragrant substances which abound in the plant world, are often hydrocarbon-like. Here hydrogen predominates. They occur there where the plant opens itself to the heat processes, especially so in flowering, or generally with plants that grow in sunny, dry surroundings.

Sugar as a Balanced Substance—Harmonizing through Oxygen

It is very exciting to study a substance like glucose in relation to the hydrocarbons. Until now we considered glucose as a carbohydrate, as a compound of carbon and water $(CH_2O)_6$. However, we can also consider glucose as substance in which the one-sidedness of carbon and hydrogen have been conquered and kept in balance by oxygen. Hydrogen, the "too-volatile," becomes water under the influence of oxygen during combustion and condenses to water and thus receives its place under the crust of the earth. Most of the water on earth is found in the terrestrial seas and oceans. The overly-rigid carbon becomes carbon dioxide under the influence of oxygen during combustion and is freed from the earth and moves about its surface. We find a sort of crossing-over taking place. These two important examples show us that oxygen not only has a regulating and activating role, but, as it were, it has also placed its extensive activities in the service of the whole. As a result of this "correction" which oxygen makes, not only are these substances brought to the places and in the states which enable them to fully be a part of the cycles of nature, but, moreover, their one-sidedness is placed in the service of life. The secret urge of carbon as carbon dioxide to re-unite with the earth and the hidden longing of hydrogen as water to rise again, attain a certain fulfillment in the plant. Together the ascending water and the descending carbon dioxide form the carbohydrates, the building material of the plants.

Not only the chemical composition of sugar, but also the way it is created point to a harmonizing of great opposites.

What we have developed here as a mental image of the overall chemical nature of sugar is intimately related to the image which the plant presents us. We also find an upward-striving, etherializing, opening-up gesture in the upper part, and a contracting, holding onto the earth, in the lower part of the plant, and in the middle the play of leaves in which oxygen takes form.

And while most of the sugar in the plant largely follows the direction of carbon and coalesces and becomes rigid, a small part rises to the flower, and is dispersed as nectar by insects into space.

Something More about the Task of Carbon, Hydrogen and Oxygen

Through all the observations presented, it is possible to acquire a clearer picture of carbon, hydrogen, and oxygen.

The life of every organism appears as an unending, streaming flow, which expands rhythmically into space and then contracts. This rhythm is especially noticeable in the higher plants if we focus our attention on the sequence: fully grown plant, seed, and growing plant.

The stream of life itself remains invisible to our senses. However, what we do perceive is a continuous emergence of forms which meet our senses, alternating with their disappearance. When the form has disappeared, then life is carried on through the seed, which has very little to offer the eye. In actual fact everything we see is already a by-product of life. Life conceals itself in or behind these appearances.

The task of carbon is shown above all by the coming-into-being, in the intruding upon the senses which occurs together with elimination from the stream of life. While life goes on from generation to generation, the individual form remains behind, rigid and isolated. And in the form we find many shapes, the organs and structures, which all are more or less rigid and separate from each other.

As soon as life has retreated into the seed, then everything that was form must dissolve and be brought back into the great, undifferentiated whole of the cosmos. *In this protest against form, in the retreat into chaos, we recognize the task of hydrogen.*

It is somewhat more difficult to recognize the task of oxygen. Oxygen certainly is mainly the bearer of life, but we always find combustion and decomposition accompanying breathing. However, if we pay attention to the opposition between life itself and the rigidifying form, then it becomes clear:

During growth life is in danger of continuously running aground in the products it produces. The task of oxygen among others, is to ensure that so much decomposition and then so much mobility come about, that life processes can go on and on and be continuously renewed. *Oxygen creates space for life in the midst of solidifying structure.*

In animals, the decomposition and catabolism go much further. While the plant mostly uses sugar as building material and thus fixes it in form, in typical animals every kind of fixation is avoided. For the most part sugar is consumed, whereby mobility emerges, just as with combustion of carbon to carbon dioxide. This combustion of sugar in the animal aids its instincts and desires and is the basis for the ability and urge to move.

Seen outwardly, what occurs in the human is the same as in the animal. Here we also find mobility as the result of the combustion of sugar. In the human the urge for movement becomes the purposeful movement of the Self. It is because of sugar that is it possible for our Self to grasp our body. The human I thus finds support in a substance, in which the urge to flee the earth and the urge to cling to the earth are balanced and thereby conquered.

It is a characteristic of our time that we use such an excess of hydrocarbons, e.g., just those substances which don't contain the oxygen function. That is typical for our time. Our social life also doesn't have the ability to reconcile and check against one another in such a way that the great contrasts form a harmonious whole. The oxygen function is, as it were, also absent here.

Sodium and Sulfur

Simple Experiments and their Consequences

A very simple experiment can make the essence of sodium visible. We take a small piece of sodium out of the protective petroleum bath, dry it with a piece of filter paper, and cut off a piece. We then see briefly the gleaming metallic surface, which directly becomes covered with a pale yellow, turbid layer. We then place the piece of sodium on a watch glass in a room that is too warm or too dry. If need be, place it in a specially prepared, humid space. The next day or at the most after two days, we will only find some water-like liquid. By attracting the moisture the piece of sodium has slowly dissolved. Here we have a substance which uses the smallest opportunity to change from the solid to the liquid state.

If we let sodium come into contact with water in a more generous way, by throwing a piece into water, then something quite different occurs. A violent reaction occurs immediately. Sodium metal, which is very soft, quickly melts, sizzles, and skips about on the water as a small silver ball. It becomes ever smaller until at last it has disappeared into the water. It disappears as quickly as possible.

In this we recognize a remarkable relationship between sodium and the water sphere.

That is certainly important. However, we can go further by researching what becomes of the piece of sodium, which we used for our experiments. Sodium hydroxide is formed at first. However, this absorbs carbon dioxide and forms sodium carbonate. *A piece of sodium left on its own does not rest until it reaches the salt state.*

In the elemental circle sulfur stands opposite to sodium. As we will still see with the other elements, that indicates opposite character and properties. We can thus do two complementary experiments. The interaction of sodium with water and that of sulfur with fire. For the comparison to be accurate we start by leaving a piece of sulfur exposed to the air – nothing happens. For sulfur the elemental state is quite normal. However, if we heat sulfur with a small flame, then a change occurs. It melts, turns bright brown and then reddish-brown, begins to vaporize noticeably, and soon burns with a small pale blue flame which spreads a pungent smell which induces coughing. The yellow piece is slowly consumed, and, in the end, we are left with the acrid-smelling air around us. Just as sodium disappears as quickly as possible into the water sphere, so we see sulfur move into the fire-air sphere.

We could also try to heat a piece of sulfur by slowly raising the temperature of the surrounding air. In that case the comparison with the experiment with sodium would be greater. It would be possible to heat the sulfur by using a heated flask or by throwing a piece sulfur into hot air. If we are able to design the experiments so that they meet the essence of the matter at hand, then we will always observe the most interesting phenomena. In any case it is a main characteristic of sulfur that (either through evaporation or through combustion) it will easily be drawn into the sphere of fiery air.

The Natural Cycle of Sodium

We will now further discuss the natural cycle of sodium–"sodium's natural history." We must, however, keep in mind that we are not discussing an element as such, but rather something that expresses itself in many compounds, processes, and matter cycles, and only with difficulty can we bring it into the elemental form and maintain it as such.

With such matter cycles we must note what the final state of the substance is, in which state total relaxation has been reached, and where it still exhibits some tension with respect to its surroundings.

We find the final state of sodium in sea water, as a salt of the sea. There it is completely "at home" and in a certain sense at rest. Sodium thus looks for a state of never-ending fluid mobility, in which it can participate in the surging, whirling and flowing movements. It finds peace and quiet in a very mobile state. In this context it is important that the sea is full of living organisms. Therefore, we will find sodium compounds especially there, where they can participate in building an environment to sustain life.

In agreement with this is the fact that we find sodium especially in our blood. Blood also exhibits a ceaseless circulating movement. Its function also serves as a basis for life.

The salt that remains after sea water has evaporated is a sort of mixture; its constituents stand in a definite weight relation to each other. Although the concentration of the salts in solution or total weight of water to salt may differ, the weight ratio of the salts among themselves does not. The concentration of the salt in blood is much lower than in sea water; the salt composition, however, is remarkably similar.[21] Although both do show differences, it is quite possible to use a solution equivalent to diluted sea water in the case of severe loss of blood. From these facts it may be seen that in the composition of sea salts and of blood salts, there is a strict, comprehensive structure.

In both sea water and in blood, sodium is the most abundant salt-forming substance present. More than other substances, we may expect it to have properties which justify this occurrence.

Sodium is not only present in water but it also contributes substantially to the minerals found in the igneous rocks. Far below the surface of the earth they are very stable. However, no sooner have they reached the surface and are they exposed to the effects of wind and weather and temperature changes, to air and water, than changes occur. These minerals are also attacked by certain organisms. Under such circumstances, sodium and potassium compounds are the weak point. If they are attacked by carbon dioxide-rich water, then they dissolve away as bicarbonates. The remaining part of the mineral disintegrates to sand-size particles, and weathers to clay. In a certain way sodium doesn't belong in igneous rock. It is being held there, as it were, in an inappropriate state, and, thus, there is some tension, especially when it comes into contact with water. It immediately starts off towards the sea.

It is apparent that what we saw in miniature in our experiment with a piece of sodium lying in the open air, is something that happens on a large scale in the crust of the earth. With its sodium-containing constituents, the whole crust of the earth has been given the drive to dissolve, to become a planar surface. If it only depended on sodium, the earth would become one large mobile drop.

In this discussion we shouldn't forget that parallel with the liquefying there is the striving to make everything into one large whole where every part is related to the whole and to each other part. An individual stone is much more of a unity; it has the tendency to isolate itself from the whole. The great currents of the ocean, and the circulation of water generally, make the earth, under the present circumstances, into a unit. In blood, which ceaselessly flows through our whole body, this unifying and binding principle also becomes even more apparent.

Properties of Sodium Compounds in Relation to its Place in Nature and in our Body

Most sodium compounds dissolve well in water, i.e. sodium gives its compounds the ability to easily join the liquid state. The solubility of water-glass (sodium silicate) is remarkable, because nearly all other silicates are insoluble. No less remarkable is the solubility of soap. Solid soap is a sodium salt of a fatty acid which is still so closely associated with the hydrocarbons, that is has water-repelling properties and mixes with oils.

Many sodium compounds are hygroscopic, i.e., they attract water from the air in which they dissolve.

From these properties we can see that sodium takes up as well as reveals its character with water.

In this context we may also view the fact that ordinary window ("soda-lime") glass contains sodium as a "flux" to increase its ability to melt easily in fabrication.

Of all sodium compounds, caustic soda (sodium hydroxide) has the greatest chemical activity. We may thus expect particularly aggressive dissolving capabilities. We may also expect capabilities which are comparable to the digestive activities of our intestines, as the juices of the intestine react fairly alkaline because of the presence of sodium and potassium compounds.

A solution of caustic soda or sodium hydroxide can also make a variety of otherwise insoluble organic substances soft and slimy and even dissolve. Furthermore, it is possible to dissolve silicic acid and fatty acids with the aid of sodium hydroxide.

The Natural History of Sulfur

Just as it is typical for sodium to be found chiefly as a salt, it is typical for sulfur, notwithstanding its great combustibility, to be deposited in elemental form.

Elemental sodium quickly becomes a compound. It will always be transformed in the sense of an energy gradient, in the direction of lessening tension. It assists the erosion of mountains. It will always seek that state which is most harmonious and has *most completely come to rest.*

Sulfur is always associated with concentrations of energy. For example, it is deposited just where mountains are being pushed up through volcanism. When we find sulfur in the elemental state, it is charged with energy which can be released when in burns.

Sulfur will be often found where a well-ordered, clear state changes into chaos. A typical example is its occurrence in mud ditches. Living organisms are always well-defined and formed. During disintegration and decomposition, the variety of forms which existed next to each other become one undefined mass. There we often find deposits of sulfur.

What about the sulfur deposits which are often connected with volcanism? We also find here a tremendous process of disorganization or chaos taking place, and even to a certain extent a violation of the established order of nature. Normally, the solid crust of the earth is that part of the nature which is most defined, least combustible, and has reached a stage of ultimate rest, and precisely here we find upheaval and movement. An incandescent mass rises and pushes the crust of the earth away or even causes it to melt. Stones, ash, and smoke are forcibly hurled up into the atmosphere. What lay deep under the crust is thrown up, and occasionally the atmosphere itself answers with thunder and lightning. There, also, all spheres are thrown into disorder: clouds rise, lightning strikes, large parts of the surface of the earth may disappear, new parts created. Volcanism is a strange mixture of destruction and creation. It seems as if primal processes of creation have been transposed into the midst of our world. Because the existing order is being breached, the way is being cleared for new conditions.

Although it is quite easy to imitate an eruption of a volcano with sulfur, it took me some time to find the best way to do this. We only need to light the top of a small mound of gunpowder to get a nice image. We can also cover the mound with some sand. However, we must, of course, take care. We get a much more impressive image if we make a mound of potassium chlorate, cover this with some sulfur, and then some sand. In the top of the "mountain" we must make a crater-like depression and in the middle put some sulfur. Everything will go well if the volcano thus formed is ignited at the top. Initially it burns slowly. However, as soon as the top layer becomes so thin that the potassium chlorate participates, the eruption begins. Particles will fly around [note:

keep quantities small – $KClO_4$ is a DANGEROUS strong oxidant and can explode!].

If we make such a heap using only sulfur, then the molten mass can flow down as a viscous liquid, just like lava.

Sulfur was called sulfur or "solferos" meaning "bearer of the sun" —a noble title. How are we to understand its meaning? Imagine how in certain areas pieces of sulfur, which often have distinct crystal forms, lie as beautiful transparent stones between others. If we then heat a sulfur-containing "stone" (as was customary in earlier times when extracting sulfur), we notice that the stone matrix hardly changes while the sulfur starts to melt, to vaporize, and even to burn. In that case heat will be evolved, and the substance extracts itself from the confines of the earth and seeks the atmosphere, where the sun holds sway. Just as the sun is the great bearer of warmth in nature, so sulfur is its bearer in the mineral region. During burning, sulfur follows the example of the sun with respect to heat radiation. It also follows the sun with respect to direction of movement and expansion. In this way we have clarified the name "sun bearer" to a certain extent.

Sulfur in Protein

If we look for sulfur in the human body, we will find it again in various forms. In its most typical and active state we find it in protein, in the protoplasm of the cells. Protoplasm is the most vital part of the organism. The more living the protein is, the more remarkable is the state in which it appears. It presents a minimum of structure which in addition is quite variable. However, from out of the slimy, undefined, mobile mass new forms, with sharply defined contours, are continuously being born. That is possible because the protein itself can change to a firmer or even a tougher state, as with the creation of horn, or change to secret other substances, as with the creation of cell walls in plants. Just as it is only slightly possible to make out clearly visible structure in protein, so also is it impossible to discover fixed chemical structure in protein.[22] Here substance is in a state of nearly total chaos. This state is so much more evident when protein is active in processes of growth and reproduction. Rudolf Steiner has pointed to these facts with great emphasis,[23] because only this opens the possibility to understand how life forces can intervene in substance and how they guide and form the processes that take place. In creating this chaos, sulfur plays a very significant role. Sulfur also stands here, as it were, at the beginning of creation. From this point of view sulfur may be regarded as the opponent of carbon, which forms the basis for the creation of solid structures.

We can view the role of sulfur in protein as a higher metamorphosis of volcanism. During the process of life the regular order is demolished, and the spheres are thoroughly mixed. What in the vital processes may be compared to the crust of the earth, that is the substance in its dead mineral state in which everything is focused on fixed chemical structures. The sphere of the vital forces, however, may be compared to the atmosphere surrounding the volcano, into which all processes are driven. With the co-operation of sulfur, substance is loosened from its physical fetters and, as it were, elevated, and placed in the realm of the vital processes. In this state the substance obeys the laws of life. Later, it will sink back into the sphere of the purely mineral-physical in a form that will be determined by the process of life. Sulfur makes substances responsive to higher influences, as it were.

Sulfur is the great transcender of barriers, in a way going up and down a ladder from one floor to the next.

The comparison with volcanism is valid in so far as in man and animal, production of protein is only possible after previous destruction. During digestion, foreign protein has to be completely decomposed before one's own protein can be created.

We can see in this that the task of sulfur is really a very noble one, however chaotic and obnoxious its many phenomena may be (smells, mainly).

A Comparison of Sodium and Sulfur Processes

By comparing the processes associated with sodium and sulfur once more, their characteristic tasks become clearly visible.

Generally, the sodium processes express themselves in the horizontal plane; they thus conform to the stratifications of the earth. They proceed according to strict laws and result in an energy gradient, as we saw previously.

Sulfur expresses itself preferably in the vertical plane and takes part in processes in which layers are overturned, order is destroyed, and energy concentrated.

In a sense, sulfur protests against the rigidifying, boundary creating tendency of the earth.

Table salt, the most typical sodium salt, crystallizes in the shape of a cube and, thus, fits very well into the rigid laws of nature, which lead to disintegration, and, thus, has always been regarded as a symbol for the earth.

The protest of sulfur really relates to the separation of spirit and matter. In our organism we find sulfur processes especially active in the abdomen. That is the first place where proteins are destroyed; there substances are carried up into the region of life; there substance and spirit work as one.

We find the sodium process above all in the head and the intestines. The crystallization of rock salt offers a good picture of what occurs in the head. We can force crystallization out of liquid as follows.

Add concentrated hydrochloric acid dropwise to a saturated sodium chloride solution, while illuminating from the side. It appears as if clouds of minute particles are glittering in the liquid. These particles grow and descend as a sort of "snow" made of small cubes. Just as substance here disengages itself from a unity of which it was part, so to a certain extent, in the head matter is loosened from the control of life processes. The released life processes can now be available for thinking. The fact that we are aware of our thoughts is due to a sort of mirroring off of matter. Consciousness is thus connected with the decomposition of what was initially constructed.

It is a remarkable fact that sodium is also involved in alkali or at any rate alkaline-reacting substances in the processes of decomposition that take place in the intestines. While in the brain we have decomposition of substances that have been synthesized by our body, in the intestines decomposition of substances foreign to the body occur. The foreign life must first be overcome before the substance can be processed. In the brain matter has to be eliminated, so that the vital forces are liberated and may be used.

It should be clear now that table salt helps to brighten a dull state of consciousness. In contrast, increasing available sulfur causes dullness, because the catabolism in the head is counteracted. People in whom the sulfur process is too strong do not sufficiently develop the ability to "precipitate" matter at the correct moment. Therefore, they find it difficult to mirror their mental pictures. They don't have intuitions at the right moment. They also have more blood than normal in the skin and have a certain propensity towards the chaotic, because in their organism the various regions, which should be separated, overlap.

Examples for Discussing Sodium in the Classroom

In the foregoing we have discussed sodium and sulfur in a more general manner and hope to have assisted the readers who want to familiarize themselves with the subject matter.

We now want to show how the same content could be discussed in the classroom. At the same time we can take this as an example of how we can move from a general understanding to a result which is worked out in great detail.

We can start by showing some important compounds. First, table salt should be mentioned here. To show it particularly vividly and characteristically, we could scatter some fine table salt on a piece of black paper or in a cylinder glass filled with water, which is illuminated from the bottom. We could also show larger crystals, which are always rectangular, cubic or form rectangular terraces. Then we can show pieces of caustic soda, which quickly become shiny because of the moisture. We can dissolve some caustic soda in water and by touching let the pupils experience that heat is evolved. We can also show how table salt dissolves and recrystallizes.

Now we take a piece of sodium, and show how light and soft it is. We slice a piece off (under oil) and show the shiny surface and then how this quickly darkens due to oxidation in air. A piece is left in a moist surrounding to be observed over the next few days. Now we throw a small piece of sodium in water and pay attention to the noise it makes and the amazing way it moves [DANGER, use a large, 1-liter, tall-form beaker to contain spattering]. If we bring a flame to the sodium, then a small yellow collar of flames will come about. We might even let a few globules of sodium perform a dance with one another. We shove a piece of sodium on a small sieve under an inverted cylinder glass and catch the evolving hydrogen gas.

It is now possible to fill a beaker glass with a solution of phenolphthalein [or first, red-cabbage] indicator. Throwing in some sodium, we will see a red color emerge indicating an alkaline reaction. The solution also feels slippery.

The next day we can start with the phenomena which could be observed on the clean piece of cut sodium. It behaves as a metal and has typical metallic properties; however, it readily throws itself into a different state. It really does seem at home in the metallic state. We can then discuss the change into an oxide through the reaction with oxygen, then the transition into a hydroxide through the reaction with water, and then the change into a salt through the reaction with an acid [CO_2].

We discuss the reaction of water with the piece of floating sodium and will analyze the process from a physical and a chemical viewpoint. Because the reaction is very violent, a lot of heat is evolved. With its low melting point, sodium easily becomes a liquid sphere. Because of its low specific gravity, it will float on water.

The following chemical changes occur:

$$\text{sodium} + \text{water} \rightarrow \text{sodium hydroxide} + \text{hydrogen gas}$$

Such a reaction is usually given as a reaction equation:

$$2\,Na + H_2O \rightarrow Na_2O + H_2(g)$$

$$Na_2O + H_2O \rightarrow 2\,NaOH\ (=Na_2O\cdot H_2O)$$

or, in summary:

$$2\,Na + 2\,H_2O \rightarrow 2\,NaOH + H_2(g)$$

This is an example of how we do not make our train of thought really conscious to ourselves. In reality only the third equation is of concern, while the first and the second are only there as explanation; in thinking we break the process up into two partial processes. Thus, we should write the equations in the following sequence:

$$2\,Na + 2\,H_2O \rightarrow 2\,NaOH + H_2(g)$$

$$\underline{2\,Na + H_2O \rightarrow Na_2O + H_2(g)}$$

$$Na_2O + H_2O \rightarrow 2\,NaOH\ (=Na_2O\cdot H_2O)$$

Or:

$$\underline{2\,Na + 2\,H_2O \rightarrow 2\,NaO + H_2(g)}$$

$$\left\{\begin{array}{l} 2\,Na + H_2O \rightarrow Na_2O + H_2(g) \\ Na_2O + H_2O \rightarrow 2\,NaOH\ (=Na_2O\cdot H_2O) \end{array}\right\}$$

After having discussed these processes accurately, then it is possible to look at them from a more imaginative way. (We could, of course, also have begun the other way around). We could discuss how sodium becomes liquid when exposed to the air or completely becomes one with water in nature and our body. That has already been previously discussed.

Of course, we discuss how caustic soda is produced industrially from rock salt. It might even be better to start with this production before discussing the properties of the metal, because that is the practical sequence.

We will save time by discussing potassium simultaneously. However, we should be careful to characterize the extent to which it distinguishes itself from sodium and stands in opposition to carbon. The phenomena of both elements are quite similar and yet differ in typical ways. We can say the same about their task in living nature.[24]

Perhaps the picture of potassium becomes clearer if we discuss it separately. We then also have the advantage of being able to recapitulate what was discussed about sodium.

An Example for a Discussion of Sulfur

First we show flowers of sulfur, as needles, and in a natural form, if possible. Then, we melt sulfur in a glass vessel. The best results are obtained if we can use a retort or a moderately large 1-liter flask. We are then able to show the various states which occur when the temperature is increased. We observe the brown vapors, and perhaps even flower of sulfur condense on the cool upper surface of the flask (actually sublime vapor to solid directly).

Something very beautiful can be seen if we place two microscope slides with some powered sulfur between on a wire screen and tripod, and gently heat them with a Bunsen burner (a hotplate is more controllable). After a short time the mass melts. As soon as it is cooled, a very beautiful crystallization process starts. So-called monoclinic sulfur appears first, which looks like needle work, shining like mother of pearl. But the next day we can observe dull yellow spots, because sulfur is slowly changing to the rhombic crystal form.

It is common to show "rubbery" sulfur. To do this, we heat a small amount of sulfur until it becomes quite fluid, and then pour it quickly before it can cool into cold water. The stream of fluid congeals immediately to a dark brown coil with a rubbery consistency. That is sulfur in an unusual amorphous state. After one or two days it also will change into the rhombic crystal state.

Here again is an example of an experiment that reveals only just a fragment of a spectacularly dramatic process. This can be seen more completely if we heated quite a large amount of sulfur in an iron (plumber's) pot until it boils and flames. When sulfur boils, we see a chaotic, bewildering, very impressive image in front of us. If we darken the room, we can also see the pale blue, mysterious sulfur flame. We now grasp the pot with a pair of tongs and carefully pour some sulfur; it flows as a blue-glowing stream onto a plate beneath and continues burning.

A large-scale phenomenon can be evoked by removing the shades and quickly pouring the boiling sulfur into a large beaker filled with cold water. Take care as the process can be so wild that an explosion may occur. If everything goes well, then we hear a loud sizzling or even shrieking. Then observe how the dark brown mass moves fiercely and spasmodically under water. Sometimes the mass surfaces a few times and subsequently collapses, until everything ends in a final convulsive movement. We see before us something that appears as a dark, dragon-like form.

According to tradition sulfur is connected to the devil. Rudolf Steiner points to a certain problematic side of sulfur, which gives occasion to speak about "dragon forces." Such things can always be made visible in an experiment to some degree. In a well thought out demonstration experiment, not only are phenomena of nature made visible, but images of the spiritual aspects of a substance can also be evoked. We can discover such phenomena by working further with anthroposophy. It is clearly not the task of the teacher to teach these points of view to the students. However, if one has developed experiment based on that viewpoint, then one should certainly show it. A profound impression of a certain side of the nature of sulfur will have been evoked. When we consider the delicate structure and the mother-of-pearl like luster of the triclinic sulfur between our microscope slides, we have the other, delicate and noble side of sulfur before us in a beautiful image.

It is a good thing to let the pupils smell the acrid, cough stimulating smell of burning sulfur. Possibly, we could show the bleaching (oxidizing) action of sulfur dioxide gas on colored plant juices. We should certainly demonstrate that this gas turns litmus red, and thus has acidic properties. We could now show in various ways how iron and sulfur react. Iron and sulfur are mixed in the proportion by weight of 56 to 32 and the resulting mixture ignited. Iron sulfide forms, accompanied by a strong glowing. Here again we have a traditional experiment by which the concept of "compound" is usually demonstrated; usually this experiment is done in a closed test tube to show that air is not involved.

If we forgo using this experiment as a proof, then we can make a very beautiful experiment in another form. We take a sufficient amount of the mixture and make an irregular, long mound. When we ignite it at one end, we will see the mass starting to glow strongly, and the glow slowly moves along the mass to the other end. Especially in the dark this presents an impressive spectacle, which is enhanced by the blue flame shot through with iron particles.

Above all we should not forget to blow some iron filings into a sulfur flame. The best way to do this is to use a large iron bowl with burning carbon disulfide, over which we sift iron filings. We then see a large pale blue flame into which the warm, orange colored iron filings shoot. This then presents an image of how the dull sulfur flame is enlightened and formed through iron (this experiment should, of course, be done in a fume hood).

The next day we will review these phenomena more closely, recall them from memory, and describe them more precisely. We will also discuss them as being expression of a being which has its specific place in the cosmos. Sulfur reacts strongly to warmth and undergoes many changes through its influence. The phenomena are somewhat surprising; they seem illogical. Partly something dark manifests itself; partly something very noble appears. We allow the overall pictorial impression of that dragon-like sulfur to have its say, but also elaborate as much as possible what occurred physically during its creation. It is then possible to find the connection to volcanism and the chaos created by warmth which is simultaneously brought about. It will then also be possible to find a link to the subtle way sulfur acts in the proteins.

Then we will focus on the process that occurs during the burning of sulfur. It is possible to derive the following reaction equation:

$$\text{sulfur} + \text{oxygen} \rightarrow \text{sulfur dioxide}$$
$$S + O_2 \rightarrow SO_2$$

Sulfur dioxide is sulfur activated by oxygen. That also explains the strong action it has to make us cough. Sulfur acts on the digestion, excites it so that the blood becomes too active in the lungs. Those of us sensitive enough will feel themselves dull in the head, because the metabolic system becomes too active there, also.

Here we encounter a dark, problematic side of sulfur: its tendency to create dull, chaotic states through its excessive activity. We can then discuss how iron in the blood has the task to maintain the balance. In pyrite, iron and sulfur have reached an ideal equilibrium. Thus, we can understand that this substance can be used as a homeopathic medicine for certain types of coughs.

When we now continue to demonstrate experiments, we can again show some of the properties of sulfuric acid—its viscosity, the evolution of heat when mixing with water, the carbonization of a piece of paper or wood, etc. The evolution of heat when mixing with water may be shown more clearly by heating water close to its boiling point, and then carefully adding a drop of concentrated sulfuric acid. Every drop causes a fierce fizzling and bubbling.

An interesting phenomenon is that zinc is not attacked by concentrated sulfuric acid, however, it is attacked by *dilute* sulfuric acid, accompanied by the fierce liberation of hydrogen gas.

A related demonstration is to show one or another acid displacements, because sulfuric acid is often used for such reactions. We can furthermore place a sulfite in a small bowl and add some acid. Sulfur dioxide will then be released accompanied by sizzling and effervescence.

It is nice to demonstrate how sulfuric acid is made using sulfur or a sulfide. The production of sulfuric acid is one of the most important processes in the chemical industry.

Then we can demonstrate the production of hydrogen sulfide gas using a sulfide and sulfuric acid. We can demonstrate how lead acetate paper turns black as an indicator for this gas. We should pay special attention to the fact that silver changes color, because under the influence of sulfur it readily forms a thin skin of silver sulfide. In general, silver is not very reactive and is an example of a noble metal. Its sensitivity to sulfur, hydrogen sulfide, and other sulfur compounds is its weak point. (If we pursue the matter further, we find that relates to the physiological task of sulfur and to the specific medicinal activity of silver in our organism, both of which are directed to the same region of our body[25]).

The exceptionally unpleasant character of hydrogen sulfide, which is consistent with its poisonous properties, can be shown by passing it through solutions of heavy metal salts, such as copper sulfate or silver nitrate. Mostly we get a very hideous, dark precipitate. If we want to produce these precipitates in a relatively large amount of fluid, so as to increase the illustrative effect and to avoid the offensive smell, then we should use sodium sulfide. Dissolve some in a large cylinder glass filled with water, and add a concentrated solution of silver nitrate. A precipitate of silver sulfide is formed, which resembles impurities in sewage water. Sodium sulfide added to a solution of iron (II) sulfate produces a very gloomy precipitate.

It is also possible to produce a typical image by passing hydrogen sulfide gas over a vertically placed filter paper wet with silver nitrate solution. A melancholy image of dirt and putrefaction emerges.

On the third day it is possible to give an overview of the three most important groups of sulfur compounds:

| hydrogen sulfide and sulfides | sulfur dioxide sulfurous acid sulfites | sulfur trioxide sulfuric acid sulfates |

We can realize the significance of these substances by looking at the soil life and generally by looking at their occurrence in nature. In Western Europe, we first come across hydrogen sulfide. In places where massive decomposition of proteins takes place, e.g. polluted canals, badly cared for compost-heaps, it can manifest itself very unpleasantly. Sulfur dioxide occurs with volcanic eruptions. It is also found in the combustion gasses from factories. Its acid rain can be disastrous, especially for conifer forests.

Hydrogen sulfide is absorbed and digested by sulfur bacteria which first release elemental sulfur, then oxidize it further to sulfates. Together with water sulfur dioxide reaches the soil. If the soil life is healthy, then this substance will also be oxidized and transformed into sulfate. Neither sulfuric acid or other very reactive chemicals are found in great quantities in nature. Nature's wise order protects us and other living-beings from sulfuric acid, in that the combustion of sulfur can never go beyond the stage of sulfur dioxide.

Plants are especially dependent on the sulfates in the soil for their sulfur intake.

If the proper soil life is disrupted, for example because of a deficient air-supply, then processes of reduction can begin, in which sulfates may be reduced to sulfites or even to sulfides and hydrogen sulfide. Sulfites are more or less unfavorable for the roots of plants, and hydrogen sulfide is very dangerous for plants.

In these three groups of substances we again have a beautiful example of the manner in which oxygen activates substances and makes them amenable to the processes of life. If we compare the formulas: H_2S, H_2SO_3, and H_2SO_4, we can see immediately that hydrogen sulfide at the most can be a very weak acid. It is sulfur set in motion be hydrogen, but not activated by oxygen. To a certain degree that explains its offensive smell and above all its exceptionally poisonous action. Hydrogen sulfide attacks the blood in its center by binding with the iron of the hemoglobin to form a sulfide.

Sulfurous acid, which contains quite an amount of oxygen, is clearly an acid, but not a very strong one. Sulfuric acid with its high oxygen content is a very strong acid. The power of sulfur, fully enhanced through the oxygen, expresses itself here.

As a conclusion, perhaps the following scheme may serve as an image of the special relationship between sulfur and heat, which was called forth for the Alchemists in their "Tria Principia" of Sal, mercurius, sulfur:

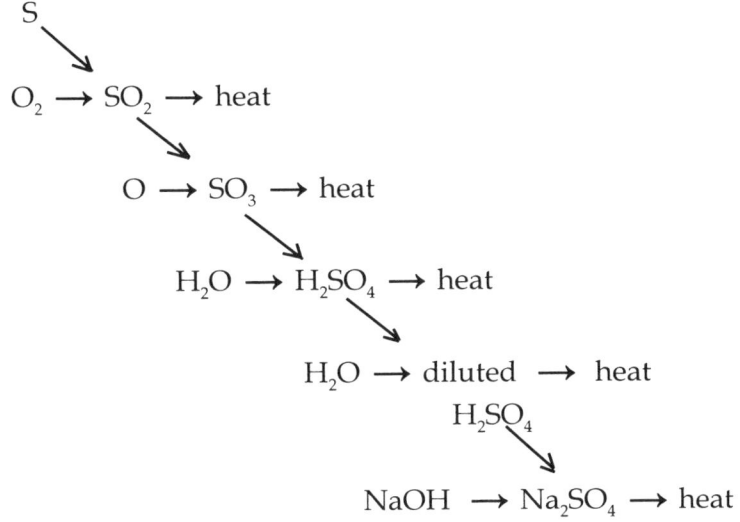

Potassium

A Comparison with Sodium

In many ways potassium shows such a great similarity to sodium that it is not at all easy to portray its own character.

In minerals, potassium compounds not only correspond with the composition of sodium compounds but also in their frequency of occurrence. The potassium compounds also serve as a point of attack for water and carbon dioxide gas.

An important difference in quantity is found if we look at their presence in sea salt. Potassium is present in much smaller amounts than sodium. Similarly, in the salts present in blood we find much less potassium than sodium.[26] Potassium thus plays a less important role, than does sodium in water, in the biosphere.

As a metal, potassium is even more active than sodium. The silver-colored surface of a piece of potassium, which has just been cut, is very rapidly covered by a bluish layer. (In the case of sodium this layer is dull yellow). If we throw a piece of potassium in water, it will react very fiercely. Immediately, a pale lavender-colored flame lights up, often followed by a crack; red sparks jerk through the flame. There is no quiet moving around accompanied by a gentle hissing, as is the case with sodium. We see that potassium also seeks water, but in a more fiery and fierce way than sodium. We could just as easily call it a fire seeker as a water seeker.

The difference in the color of the flame is also meaningful. Despite the wildness of the potassium flame in the experiment above, the color of the flame has something fragile and delicate. In contrast, the sodium flame has a conspicuous yellow color. In other respects its appearance is very unassuming – it only shows itself when sodium is ignited, and it surrounds sodium as with a yellow, closely-fitted collar. Sodium shows a certain restraint where fire is concerned, in contrast to potassium.

As is common, such differences between the nature of these substances become clearly apparent in their technical use. Potassium salts, such as potassium nitrate and potassium chlorate, are used in gun powder, fireworks, matches, and generally in the pyrotechnic industry. Sodium salts are not used, even though they are cheaper. Potassium salts are less hygroscopic than sodium salts.

Only when comparing the roles of sodium and potassium in the living world does the typical nature of potassium compounds become apparent. Plants that grow on land must take their nutrients mainly from the soil as soluble salts. Predominantly, they choose potassium salts. Thus, potassium is the most important mediator between their roots and the mineral world.

The only salt which is absorbed in pure mineral form by animals and the human body is table salt, which is a sodium salt. A potassium salt absorbed in the same manner would act as a poison. As previously mentioned, sodium salts predominate in the extra-cellular fluids of animal and human tissues.

Seaweed, which lives in an environment rich in sodium salts, shows more of a resemblance to animals with regard to the salts in its body fluids than to land plants; its ash contains mostly soda or sodium carbonate, while that of land-plants contains mostly potash or potassium carbonate.

A comparison of the blood circulation in animal and human with the movement of fluid in the plant is highly interesting and undoubtedly very significant for the character of sodium and potassium. Blood moves in a rapid, repeating, circular movement, although not in a completely closed circuit, since water or salts are continuously being excreted, and salt and water are continuously being absorbed.

While blood not only circulates steadily but always moves rhythmically to the center and away again, in the plant water rises in a rhythm which is determined by day and night; water is on its way from earth to sun. A small part remains behind and is used by the plant. The greatest part flows into the leaves where it evaporates into the atmosphere. The salts which have traveled with the water are partly used to produce plant substance and partly remain as salts. It is possible to show an increasing salt content of the plant as it grows older. That is especially apparent in the leaves, because these have a higher ash content than stem and roots. So by dropping leaves, trees return large amounts of minerals to the soil.

A Comparison with Carbon

We can get a characteristic idea of the task of potassium by noticing how the plant gives structure and form to water.

In land plants we will always find a skin or at least a boundary layer which is impermeable to water, which surrounds everything. Within that is mostly water-rich tissue wherein water may move in all directions. We find the vascular tissue nearer the center, in the stem. These consist of fairly rigid, nearly impermeable walls through which water flows. In the living cell we mostly find a space in the middle which only contains water and substances dissolved therein, surrounded by a zone of watery protein, and then a more rigid, less water-containing cell wall. We always find a more or less mobile mass of water which is enclosed by a wall. We, thus, create an image of the plant as a fountain of water, which is caught, enclosed, and partly held. This retaining and enclosing gives rise to a second flow, which chiefly moves down and carries down the substances through which the plant builds itself up. Those are primarily products of photosynthesis and other substances, mainly grouped around carbon. The salts in the first flow are chiefly grouped around potassium.

Thus, we find a flow typifying potassium, which rises from out of the earth and disperses in all directions, and another one, characteristic of carbon, which flows down, surrounding the first, giving it form and partially bringing it to rigidification. When water predominates too much, as is the case in moist, warm, shadowed places, plants acquire the propensity to swell and run rampant, and, thus, become soft and limp. If the second stream becomes too strong, as is the case, for example, in very bright and dry places, then the plants acquire a compact and solid form. In the first case too much water remains in the plant, which is not contained enough. In the second case too much containment and too little water are present. In a normal plant the rigidity of the structure comes about through a balance between both flows of substance. The upward flow leaves its salt and part of the water behind. The concentrated solution which results is taken up in the structures which are excreted by the downward flow and to a certain degree sealed up.

It appears as if the ascending salts which remain dissolved to a large extent, entice the products of the photosynthesis to harden and to condense, in short to move towards mineralization.

Also in our body potassium is found more within closed structures than sodium. Potassium salts are found more frequently inside the cells, while sodium salts are more common in the extra-cellular surrounding fluids.

Silicon and Silica

The Occurrence of Silica in the Crust of the Earth

More than most other substances, silicon presents its essence in an image. This image is that of the quartz crystal. It consists of silicon dioxide (SiO_2). Although generally we have to study dynamic processes to begin to understand the nature of a substance, here the essence expresses itself in a static image.

However, to penetrate into the depth of this immobility, we ourselves must become just so much more mobile. Doesn't this purity deeply move us? The substance is hard, impenetrable, and excludes us from its sector of space, but still invites us to enter into a glorious building, as it were. There we are surrounded by a cool beauty. Along with light, we may penetrate into every niche and cranny. Sometimes we encounter an imperfection, a crack, or a small cloud. But these things seem to be there only to enhance the beauty and play with the light. The small clouds shine with an unearthly clarity; every fissure shows a color-play more beautiful than we can dream of. At times, such a crystal resembles a house with many rooms. And every room, no matter if large or small, has a similar shape. It is always longer in one direction and narrower in the perpendicular dimension. The walls are built like the cells of a honeycomb; thus, a six-sided column arises with a pyramid-shaped roof.

Gravity had nothing to do with making this form; it is impossible to tell how a crystal was orientated as it grew. However, it is quite possible for the earth to impede the growth of the crystal. As a consequence, we hardly ever find crystals which have developed the pyramid at both ends of the crystal. They generally have grown sticking out from a solid, earthy base.

But precisely this interruption at the base of their wonderful structure gives the appearance of indicating something with a mighty gesture. In their material nature, they are connected with the parent rock; through their form they point to the farthest reaches beyond earth. When we open such a cavity in a seemingly barren rock to discover sparkling crystals pointing from all directions towards the center, then we see that silica's growth gesture is in all directions. The crystals are part of the earth. They are supported by the parent rock, but these crystals, which are composed of this most characteristic earthly substance, show us precisely that the earth is a transparent body. Whoever really identifies themselves with such a cluster of crystals feels on all sides and under foot the abyss; they feel themselves as if they were freely floating.

It is no wonder that with this transparent substance the most delicate, pure, and transparent colors can appear, especially the violet of amethyst or the ethereal pink of rose quartz. Now and again, we find the indication of threatening dark, as in the black of smoky quartz—but it too remains semi-transparent.

Quartz appears to be made from a substance that seems to have been created for a never-ending play with light. It is among the most beautiful crystals found in cavities deep in the darkness of the earth's crust. It could happen that during the drilling for a large tunnel, the wall broke open. By approaching with lamps and stepping into the dark cavity behind the wall, one found oneself in a rock chamber with mighty shapes, which stuck out of the wall towards the center. Huge crystals, meters high, gleamed in the darkness and made the torch light dance in their interior. These rigid, precise shapes have rested this way in the earth's depths in total darkness and peaceful immutability. In addition, countless crystals also rest, still in their undiscovered cavities.

By comparing this with the ever-changing surface of the earth, caused by wind, clouds, and water, by heat and cold, dark and light, we can then get a strong impression of the essence of the rocky crust of the earth. It rests as a memorial to a far past, when the most violent changes took place, where everything that is now rigid was still taking part in the processes of life.

When considering silica, it really is justified to think of the essence of the crust of the earth, since a large part, at least 97%, consists of silicates and silica. The properties of the crust of the earth are determined in large part by silica, and we can consider the quartz crystal as an ideal crust of the earth.[27]

While that titanic force in the interior of the earth, which erected such heaven-seeking even heaven storming constructions, has come to rest, we find a continuous disintegration taking place on the surface. Because of the mutual effect of heat and cold, through the infiltration of carbon dioxide-containing water, rock disintegrates. Part is dissolved and carried off to sea; part weathers into a fine sediment which is carried along and deposited elsewhere as clay; part is ground to sand which is quickly deposited.

If we investigate still deeper layers of the earth, then we again find a region where everything is in motion. There, warmth is active and keeps everything in a viscous state. This movement, however, is not directed only towards decomposition, but also towards creating new minerals.

We, thus, learn to understand the crust of the earth as a layer at rest, whose forms are attacked continuously from the outside, while on the inside new species are formed.

The Properties of Quartz in Relation to the Different Realms of Nature

If we want to make a complete picture of the properties of a substance or of a living being for ourselves, then it is a good idea to seek out its characteristic reactions to influences from the various realms of nature. Otherwise, it is very easy to overlook certain important properties.

Of course, we can not discuss this method every time we use it. However, in discussing quartz, we will use this method in detail.

Let us first move through the sphere of the elements and then through those of the higher forces of nature.

Quartz is very hard, which means nothing more than that it resists the pure mechanical influences of the earth element to a high degree.

Quartz is essentially insoluble in hot and cold water. The influence of the water element thus also has no effect.

Quartz is hardly volatile and thus shows no affinity to the air element.

The expansion due to heating is very small (the expansion coefficient is 1/400th that of ordinary glass). The melting point is very high (1713°C, glass about 1200°C).[28] These facts indicate that the warmth element has very little influence on quartz.

We continue now with light and thus follow the method Steiner indicated when he discussed the four elements and the "ethereal forces." In its pure form, silica is particularly transparent to light. In comparison with other substances, its transparency to ultra violet light is very remarkable. That means that light can pass through quartz without an interaction with quartz taking place.

There is hardly a substance that can attack quartz at normal temperatures. It is scarcely susceptible to chemical influences. If it does react under certain conditions, it behaves as an acid. There is one exception to the rule: the specific action of hydrogen fluoride on quartz. This will be discussed more thoroughly later on.

The life processes cannot, of course, take up quartz in this form.

Reviewing the above makes a good basis for understanding why Steiner could describe quartz as "something tremendously noble, that doesn't want to do anything." [29]

Silica in Living Organisms

Reflecting on the above makes it all the more remarkable that living organisms can use silica so extensively to their own ends. In humans and animals, we find an exceptionally high content of silica in hair. In birds a lot of silica is found in the feathers. This means that silicic acid is part of a process in which substance is continually being moved outward and excreted. It is comparable to a slow movement out in all directions. Of course, at first silicic acid was part of the living substance and took part in the processes in the living organism, but then it moves to less living parts of the organism, which form a sort of dried up and withered cover.

In connective tissue, silicic acid also comes somewhat to the fore. That implies that the organs, as the centers of life in the organism, shove silicic acid to the periphery. It also participates in surrounding the organs and enables a kind of perception of the opposite side.[30]

It is possible to process quartz in such a way that it has a specific healing effect on the sense organs and especially the eyes. It is not so difficult to understand that there is a relationship between silica and our sense organs. The task of our skin is to envelop us and to shield us to a certain degree from influences from the surroundings. We can see the protection from the environment as a main task of the skin. The remarkable thing is that the more outward-directed sense organs are more or less embedded in the skin. These sense organs, thus, contribute to shielding and protecting the body from the environment, but on the other hand, they are open to very specific influences. From this point of view the eye is typical. Defensive strength combined with transparency is just as characteristic for the sense organs as for the quartz crystal. We can also see here a clear relationship between the role a substance plays in our body and the way it is used in industry. Strength and transparency of a window pane, for example, are mainly due to silica. It acts as a piece of wall as it keeps the environment out, and it acts in an opposite way by being transparent. A window pane relates in the same way to a wall as the skin covering the outside of the eye relate to the normal skin.

In complete agreement with the above is the fact that silicic acid is present in plants and is found mostly in the tissues that make up the cell walls.[31] Here, also, substance is deposited from the living cell, in conjunction with being moved outward during the building of the cell walls.

There are plants and animals in which silicic acid plays such a dominant part that we may call them "silica organisms."

Primarily, we should mention the radiolarians. For the unaided eye, these are hardly visible, single-celled minute animals, which float in vast numbers in the oceans. Here, silicic acid is made into a delicate, precisely-formed skeleton. The silicic acid is secreted from a soft protoplasm mass. This group of animals seems to devote itself to the question of how many different, harmonious, wonderful shapes it can make by playing with extremely thin needles, minute rods, and perforated small sheets.

Other silica organisms are the diatoms and the silica algae, single-celled plants which float in vast numbers in the highest, light-filled layers of the oceans. The living contents of every diatom is enveloped in a silica sheath which is largely closed to the environment. This armor again consists of two parts which fit like bottom and top of a box. The structure and especially the design on the sheath or "test" is remarkably complicated and follows a precise pattern.

These facts give us a very vivid picture of life in the oceans. On the whole the plant substance of the oceans is encased within minute silica bodies, which float in the topmost zones where they can sparkle in the light. Silica particles continuously sink through the water realm. The diatoms act like many small centers in free space, towards which substances and forces are gathered together from the nearby surroundings and from the universe. The animal life in the oceans is engaged in this play of contrasting forces, because in the end all sea animals, even the largest wild beasts of prey, are dependent on the diatoms, which are the primary food for small marine animals.

On land, in many respects the situation is just the reverse. Land plants grow in soil that mostly consists of silica and silicate masses formed in the past through lithification, and they grow upwards, away from the soil. Silicic acid is found within the plants. Partly they are centers, like the diatoms, where forces and substances are gathered together from the surrounding space. However, in contrast to the diatoms, they also offer the cosmos substance from the soil and their own organism. The land animals can live because of the plants; they live more or less "on top of" the plants.

It has been calculated that all the diatoms together form a mass which is larger than that of all plants, animals, and human beings. At any rate, a large portion of the plant substance of the whole earth is enveloped in silica. From this and other findings we can see that there is a special relationship between silicic acid and plants, just as there is between lime or calcium and animals—as we shall discuss in detail later on.

We will find the same difference in relationship with regard to chalk between higher and lower animals as we found in higher and lower plants with regard to silicic acid. In both plants and animals the higher organism carries within itself that which forms the outer cover in the lower organism.

There are also higher plants, especially horsetails (Equisetum) and the grasses, which we could not call silica organisms in the same sense as the diatoms, but which do contain a high concentration of silica in their cell walls. It is a remarkable phenomenon that in both these groups the high silica content parallels the tendency towards a linear structure. The grasses especially show an unsurpassed stretching vertically. In relation to their thin stems, and especially with respect to the light weight of their stems, they are the tallest organisms. Nature demonstrates emphatically that we are dealing here with an architecture that opposes gravity, by allowing such a slender structure to be crowned by a rather heavy seed-head at the top. The typical plants of this group give evidence with every feature of their structure, also in their leaves, of a direction away from earth towards the heavens.

In the discussion of quartz crystals we pointed out that these negate the earth if we look at their orientation. We will see later that therein lies a certain polarity to chalk. With respect to the diatoms and the radiolarians, we have now determined that the lower silica organisms live floating freely and do not have a specific orientation, as if gravity did not exist. In the case of the higher silica organisms, the grasses, an unsurpassed virtuosity is developed in overcoming gravity. It thus emerges again and again that this most abundant substance in the earth participates in formations and processes by which the typical force of the earth is overcome and negated. Again, we must contrast this behavior with respect to gravity with the behavior with respect to light. Just as the diatoms are unaffected by gravity, they are dependent on light. The grasses, which reach towards the sky, are also directed towards the light. Because marine diatoms occur in such vast numbers, we can say: A very large part of light that is absorbed and processed by plants has to penetrate a silica cover. We could be inclined to compare such organisms with inverted lamps; in a lamp the light source is held within a transparent sheath, and the surrounding dark space receives the light. Seen from the outside, a diatom has covered itself with a silica sheath. Seen from the inside, this being offers silica to the light. What is very apparent in the diatoms also takes place in other plants, but to a lesser degree: silicic acid is pushed outward to the walls of the cells and to the surroundings, thus always in opposition to the influence of light.

It is generally true that plants growing on the bank of a river, e.g., bulrushes or reeds, have a high silicic acid content. This whole group of plants demonstrates through their vertical form, by the upright stance of their sword-like leaves, that they have little affinity for water. Water encourages the growth of more or less wildly growing, rounded leaves.

In their posture, the reeds and rushes counter the horizontal water surface, and in every detail of their form they reject its tendency towards soft, rounded forms. But these plants not only fight against water as expressed in their structure—they also contribute substantially to the creation of land from the ditches and swamps. Thus, they cause a transition from the moist to the dry. That is really interesting, because also in our body, and in that of animals, silicic acid is engaged in processes wherein dampness is transformed to dryness. We find this taking place in those parts of the body where skin is being moved outwards and slowly flakes off or where hair and feathers grow. As we will see when discussing lime, we will find that chalk is deposited in the skeleton, a transition from the liquid to the solid. However, the skeleton remains involved to quite a high degree in the processes of circulation and life. In contrast, the rejected pieces of skin, hair, and feathers, dry up completely. Thus, silicic acid is active here in the most intense processes of death and decomposition.

From this relationship between silicic acid and the transition from water to land, we can deduce something important. Silicic acid was engaged in monumental processes of solidification and hardening during the formation of the earth. What took place on a tremendous scale in the past is now repeated in each living organism on a small scale.

To conclude, we would like to return to the silica in the feathers of birds, because there we find a sort of summary of everything we have been discussing up till now: the transition from moist to dryness, the delicate structure of lines and points, and the use to overcome gravity. If we see a large bird glide, e.g., an eagle, without a single beat of its wings, then we may take this as the most magnificent expression of the essence of silica. Not only is the animal carried as if there were no gravity, but, through its sharp eyes, it also lives in the vastness of light-filled space. By the gesture of its wing-movement, through the construction of its feathers, this animal points to the periphery of the world, to that part of the universe where light belongs and which is opposed to gravity.

The conquering of gravity and being transparent for light are the central themes of the forces of quartz.

Silicic Acid and Water

From the fact that living beings process such vast amounts of silica, it follows that silica must be present not only in the earth's crust but also to a large extent in the waters of the earth. And as a matter of fact, the whole water layer of the earth is permeated with silica in a peculiar, finely divided state, which we will discuss in a moment. Silica is continuously being gathered by living organisms and solidified from out of a state of mobility into the graceful small and large constructions of the plant and animal world. By far the largest part is given back to the earth (radiolarian-mud and diatomaceous-earth). Decomposing organic substance also liberates silica from the water.

To get a mental picture of the interaction between silica and water, we can do some experiments with water glass or sodium silicate. For example, we can pour a concentrated acid into the viscous, syrup-like solution. Immediately, an opalescent jelly is produced, consisting of hydrated silica. If we take this jelly out of the water, it will dry out to a very compact, brittle mass. If we do not let it dry out too far, it will be able to take up water again. If we rinse the silica gel properly, and leave it in water for a long time, then by a process of gradual congealing we will even get an opal-like substance. Thus, we can see that there exists a special affinity between water and silica. This does not go so far that silica dissolves readily in water. The viscosity of a water glass solution already indicates that silica, even when combined with a solution of sodium hydroxide, still retains its tendency to coalesce into a gel – it remains in the colloidal state.

If we greatly dilute water glass and then add an acid, we do not observe any phenomena of coagulation. And yet small silica particles are, as a matter of fact, being formed, and can even be seen through a microscope; they, however, are so far apart that they hardly influence the viscosity of the water. In this case we speak of a "sol." That is the state in which we find silica in natural waters.

The gel-state, which we discussed previously, is also very important. A considerable proportion of the crust of the earth consists of silica minerals such as opals and flint-stone, which have a higher or lower water content.

It is possible to liquefy quartz with the aid of over-heated steam (200°C under excess pressure). This is also of great importance to the silica circulation in nature. At many hot water springs we find large silica deposits. We find beautiful examples of this on Iceland.

It is of great significance that all silicon compounds of the earth, which tend to change to silicon dioxide, are found there where the fire element has receded and water comes to the fore. It follows from this that silicic acid when exposed at the earth's surface tends to retreat from every chemical reaction. Even with sodium silicate solution, we should not speak of a silicate, but rather of a colloidal solution of silicic acid in water which has been made alkaline. This remarkable property of silica is one of the points of attack for chemical weathering of minerals.

Silicic Acid and Warmth

To activate silicon dioxide we must always use fire and even quite a lot of heat. A very high temperature (1,600°C) is necessary to soften it up, and only at 1713°C it becomes fluid. It can then be poured and blown like glass. At a much lower temperature, but still quite high, it can be used in the preparation of glass. At around 1200°C a mixture of sand, soda and limestone melts. We get a peculiar clear and quite a thin (non-viscous) substance. There is no particular temperature at which it solidifies, but with decreasing temperature the substance slowly becomes more and more viscous. At around 800°C it becomes so viscous that it can be worked upon. At room temperature the solidification process has not come to an end, but the viscosity of glass has become so great, that it appears has a solid. However, in the case of large shop windows we must reckon with slow changes of its form, because the mass of glass very slowly flows downwards. Silicic acid thus conveys to glass its jelly-like nature. Glass only very slowly becomes a solid. It becomes milk-like, opaque, and shows a crystalline structure - it "de-vitrifies."

Waterglass or sodium silicate is also produced by using high temperatures. It is produced by mixing sand and soda (sodium carbonate). However, not only in producing waterglass do we need high temperatures, but even to dissolve it in water we need heat. We need temperatures which are far above 100°C.

A remarkable property of glass is to take on colors when we add metal oxides or even metals. That is only possible when glass is liquid at a high temperature. The metal oxides will form silicates; however, we find that just as with the metals, they are present partly as colloids. In this manner gold and silver especially are cable of producing beautiful colors. Without doubt this property of glass stands in relation to the nature of silicic acid. It lets everything, which is related to light, present itself fully.

Silica as Oxide

After everything we have said with respect to oxygen, it is not easy, in the case of silicic acid, to see the chemical composition (SiO_2) and its properties in perfect harmony. In principle combining with oxygen indicates an activation and becoming susceptible to interactions with other substances. Indeed, a great number of compounds can come about between silicic acid and other substances, which often are very complicated in structure. The composition of many minerals is abundant proof of that. Under "normal" circumstances there is little evidence of any activity of silicic acid. Both oxygen and silicon appear to be in the most relaxed, well-balanced state. By far the largest oxygen reserve of the earth is fixed as silicates or silicic acid and, thus, not in the active state of atmospheric oxygen. The element, which distinguishes itself through its ability to activate other substances, is, to a large extent, itself fixed in a highly inactive form. Does this suggest a weak point in our whole discussion?

We can arrive at an answer to this question by studying the most important properties of silicon as a chemical element and comparing these with those of quartz. As a crystalline substance silicon is so hard that we can use it to cut glass. It is difficult to melt and to evaporate. Herein properties related to those of quartz are announced. However, silicon also has properties which are almost opposed to those of quartz. Silicon is nearly black and has a metallic sheen. It is opaque and is a good conductor of electricity. It thus exhibits pronounced metallic properties, and it is related to that group of substances that totally reflect light.[32] The structure of crystalline silicon resembles that of diamond, while its outer appearance resembles more that of graphite.

Thus, the transition from silicon to quartz is a very profound one. In quartz, all the dark, earthy characteristics have disappeared and become their opposites. Quartz is very transparent for light. Here we can again recognize the action of oxygen.

In silica there lives a clear urge to turn away from the earth and the universe and to seek a connection with the dark side of the earth. Oxygen, however, has retained the openness for light. It has let itself be fixed through the powers of silicon and banned as quartz, but simultaneously it has overcome its all too great one-sidedness and transformed them in a kinship with light.

We do not need to delve very deep to see that the nature of oxygen reaches a certain culmination in quartz. We only need to see the quartz crystals jut out of the more or less formless, dull parent rock, with their majestic forms and their unique transparency for light. We are tempted to compare the relationship between the ordinary rock and the quartz crystals, with that between our everyday thoughts and those which pertain to high ideals. As they open themselves for the sunlight, so should our ideals follows lofty spiritual laws. In the same manner they should be irradiated by spiritual light. There is not much more that can be achieved on earth, but overcome the dark resistances from lower spheres, and be open for the light from above.

If we remind ourselves once more that oxygen is "sun-sent," and continues the sun's activity in the realm of matter, then it becomes clear why oxygen comes to complete rest in quartz: it can not achieve anything higher than to create earthy substance, wherein the way is completely open for the sun.

Quartz lets light pass with little resistance, and, thus, is passive, but it also easily takes on colors given by other substances, and retains them strongly. In solid silicic acid resides the power and solidity of silicon, but it lets something else come to its own, and places itself at its disposal. In these latter properties, silicic acid has a remarkable correspondence with water.

Silicon and Fluorine

It is very surprising what occurs when moist hydrogen fluoride reacts with silicates. A volatile compound silicon fluoride, SiF_4, is formed.

If gaseous fluorine reacts with silicon, the same compound results, accompanied by a rain of fiery sparks. Two elements, which are more common rock-forming elements than any others, react to form a *volatile* compound! Apparently, if we added a solution of hydrogen fluoride to silicic acid (sand) in a bowl, then the whole content of the bowl would evaporate. Fluorine has a special affinity for silicon and reacts with it in nearly all silicon compounds. At room temperature, quartz does not appear to react with fluorine readily but only with the addition of moisture and elevated temperatures. At any rate, fluorine has the tendency to transform the typical silicon process into its opposite. Not only does it take silicon away from oxygen, but it also severs the connection with the earth element. Such a process seems like an attack on the foundations of earth.

The fact that fluorine liberates oxygen from all kinds of oxides, and thus returns it to its initial active state, must undoubtedly be seen in the same context.

Silica Technology

We have often seen that technology can be read like a book of nature and of the human being. Reading not only reveals simple laws of nature, but often also real mysterious relationships.

We find a world wisdom reflected to a much larger extent in technology than with contemporary science. Very often we get the impression that in engineering we have imitated nature or a function of the human body. However, this is seldom the case. In our present discussion of silica technology we will find a whole raft of evidence for the correctness of this opinion.

Silica and silicates are present in far larger amounts in the crust of the earth than any other substance. The technology of building consists of stacking up large amounts of materials upon one another. Naturally, silica and silicates are utilized in the form of sand, natural rocks, bricks, cement, and the like.

Everywhere on earth there are sheets of water, large seas, and lakes embedded in the landscape. Rivers flow in their banks. Basically, all this water is carried and surrounded again by silica and silicates. When we use earthen bowls or other vessels made from glass or porcelain to contain liquids or to pour from, we unconsciously are imitating nature.

To fire pottery or bake bricks, or to cast glass and quartz, we need a powerful fire. That is a repetition of processes which shaped and still form the crust of the earth.

We are particularly dependent on glass when we want to make something strong and yet transparent. One of the most striking examples are window panes. We have already mentioned that the panes are placed in the walls, like our eyes in our skin.

Basically, a light bulb consists of a glass or quartz envelope and a light source inside it.

Glass and quartz can not only be used to transmit light, but also to form and direct it. Among other things, lenses and prisms are used for this purpose. These are used in appliances such as lighthouses, film projectors, etc., in which light is directed outwards, and in others like cameras and telescopes and the like, where light is gathered and directed inwards.

We have already mentioned that there is a relation between silica and our eyes.[33] At this point in our discussion this again becomes important. Sometimes we are told that our eye is like a camera. This statement is correct if we turn it around–the camera is a fairly comprehensive imitation of our eye. The same applies more or less to many optical instruments. Most of them are used to replace the function of the eye, to enhance or to aid it, as for example, glasses, telescopes, microscopes, and even lamps.

In all these cases, silica-containing substances are used for the lenses and the other transparent parts. Thus, again technology furnishes evidence for the kinship between silica and our eye.

It is very interesting that we can just as well compare a projector or a car light bulb to the eye as to a camera. Our eyes do not only passively receive light, but there is also a certain activity emanating from them: the gaze. This corresponds to the luminous organs of deep sea cuttlefish, which are built as eyes.

We can arrive at a new understanding of these applications of silica, if we realize that in the realm of light we are dealing exclusively with images. Even a light source is nothing but an image for our eyes. This is best seen by noticing the changes in size brought about by distance, or the shifts brought about by mirroring or refraction. From this point of view, a beam of light is that [cylindrical] region of space in which we can see the light source. We could also say: ...in which we can make an image of the light.

We know that our eye receives the environment as images and projects them in miniature on the retina. We can also speak about how windows are places in the walls which are transparent – both for the images from the world, as well as for our eyes to look out.[34]

In this way we understand manipulating images as the main activity of silica technology. In many cases we are dealing with images which shine through a transparent wall, as in the case of a shop window, a display case, or an aquarium.

The lens of a projector projects an image outwards (on a screen), that of a camera inwards (on a film). Images are shifted many times and enlarged or made small in a microscope or a telescope.

Thus, technology makes it clear that there is a special relationship between the nature of silica and the appearance and formation of images. (In a different way the same applies to silver.)[35]

We can achieve particularly beautiful phenomena when we create a rainbow using a powerful lamp such as a carbon arc and a carbon disulfide prism, and moving a quartz crystal through the rainbow on the other side of the prism from the lamp. The crystal successively takes on all colors, and in such a manner, that it seems as if they radiate from the inside of the crystal. That points to something we could call the "potential" of quartz. It seems to be destined to take on all colors and to radiate from the center.

And actually, nearly all artificial lighting is based on silica containing substances, from the center of which shines a light. In the case of electrical bulbs made from frosted glass, we only see glowing glass.

With the colored windows a certain culmination point has been reached. Here we find glass in all colors, and, furthermore, it is as if it radiates from the center. When the first colored church windows were used in Europe, houses did not have windows. Windows only had religious meaning. And, if we study the way they were made, we find even this had something of a religious nature. To the half-molten glass, metals or metal oxides were added. Initially, these substances were totally opaque and partly earth like. Afterwards they gave to the glass the most beautiful, transparent colors. The dark nature of earth had disappeared, and was made transparent for the light. Putting innumerable pieces of colored glass together to produce religious images could now begin, as well as setting the windows up and putting them in their place. Substance is lifted heaven-wards. We have here a sacrificial act in the realm of building. However, not only the metals come to a higher expression here, but also silica itself. Everything comes together here: taking on color, the raying out, and conquering heaviness. While in other cases, silica-containing substances transmit images, here glass becomes itself the carrier of the images. Glass shines in sun light, and the images also picture heavenly things.[36]

Silica and the Sun

From the phenomena we have discussed, the deep relationship of all silica-containing substances both with sun and earth appears again and again. Normally, we experience earth and sun as opposites. The sun appears as a free-floating heavenly form, which continuously sparkles with light. In contrast, the earth appears as dark, opaque, light-avoiding ground under us. We stand up above it and support ourselves on it as on a solid impenetrable foundation. What is silica then from this point of view, when it appears in its highest natural form as quartz or when it appears as colorless or colored glass brought to a new culmination through technology? It is earth without its ability to ward off light, an earthly substance that has given up one of the most pronounced differences to the sun. It is earth through which light has free play. Silica belongs more than any other substance to the solid part of the earth, but as quartz shares least of all in the darkness of earth. Quartz tells us: in the earth there rests a deep kinship to the sun. Earth matter at its peak is prepared to embrace the sun. If, furthermore, technology steps forward, as in the case of the church windows, then something can come about which we would want to call a solemn marriage of earth substance with the sun.

After these insights have been gained, we should reconsider phenomena previously discussed in this section.[37]

Lime (Chalk, Limestone)

Formation and Erosion of Limestone Mountains

Due to its layered structure, limestone is like an immense book. By leafing through this book, we can read the history of a geological period encoded in a sort of sign language. The deeper we go the further we advance into the past of the earth, and by looking in the uppermost layers, we find ever more signs of a period nearer to our times. In many places it looks as if the book has not been kept very well. It was folded and broken then, and often terribly crumbled and torn. But again, that is of importance to those who read seriously. In every distortion, we find a symbol of a second script which testifies to the history of the book. Moreover, we should be thankful that pages are sometimes over turned, because much of recent history, which otherwise would have been destroyed, has, therefore, been preserved. The primary script is the traces of animals. For example, within the grayish-blue mass of a limestone ledge, we can discover whitish shell-forms or skeletal figures. Our own letters are abstract signs which only relate to concepts and words in a fairly arbitrary way. These "stone figures" are real remains which not only point to previous animals, but also reveal their forms and have often retained some of their substance. We find clear remains of shells, mussels, snail-shells, but also the amorphous foundation consists in large part of crumbs and pieces of animal parts. And this infinite number of recollections is strongly cemented together and is used in building the largest buildings on earth. The nature of the second script is different. It speaks of movement and enormous upheavals which passed through the motionless, cemented mass.

At first glance, it often doesn't seem as if limestone mountains were originally made from layers that were horizontal. They often stand before us with blocky, heaven reaching summits, with mighty protrusions, and again receding masses of stone with wildly jagged edges. We can still recognize the layers, but they have been pushed up, broken and displaced, turned over and folded; it is as if they have been rolled out and kneaded like viscous dough. Everything shown by such an imposing limestone mountain gives the impression of being wild, dramatic, and strange. Wild is the reminder of the forces that are active in the depth of the earth, and strange the traces of the eroding activity of wind and weather. Deep crevices have emerged which allow large blocks of rock to go crashing into the depths. Here and there are slopes, which, as if there had been a rain of spherical stones, are pelted with fallen boulders. And water burrows and works into the landscape and is generally busy forming gullies and eroding their walls.

Nowadays there is quiet in the heart of the mountains, but changing influences penetrate from the surface inwards. Together, heat and cold, wind and water break the proud resistance of the mountain. Water trickles and seeps even into the smallest cracks, and upon freezing, breaks open the walls a little. That is how stones are split; the fragments then await the spring thaw and then roll down the slope.

From the atmosphere, but especially from the ground, some carbon dioxide dissolves in water and gives it the ability to etch away a layer of limestone from the rocks and, dissolved as calcium bi-carbonate, take it away and let it form deposits elsewhere. Through cracks and narrow veins, it reaches inner parts and can form a gleaming layer along the walls of caves, or at a particular spot it can drip continuously year after year. Often it takes quite some time for the next drop to fall. In the meantime, some of the carbon dioxide gas can evaporate; to the degree this happens, part of the calcium carbonate turns from solution to stone. Although the layer left by each drop is extremely thin, during the long years it grows into chalk cones along the path of the drops. Beneath it, the drops splash down and flow away, also forming a film. And again, layer upon layer is formed, albeit somewhat quicker then above and over a larger area. In that manner, a mighty pillar arises. This process of flowing, dropping, splashing through hundreds of years, makes the cone point grow downwards and the pillar up. In the end something like a complete pillar stands in front of us, apparently destined to carry the vault of the cave. However, nature never gives things their function in this manner. It seems to be more the case that nature uses every opportunity to play. And here, in the case of the formation of stalactites, nature plays in a manner which could not be surpassed by the most audacious imagination. It seems as if it could not rest until it had enticed all the variations from this simple theme. Wandering through these caves is like encountering castles, drapes, or organ pipes, remarkable shapes and a multitude of other forms. However we may see these limestone mountains, on a large or small scale, the rich variety of the play of forms draws our attention.

But, however extravagant the play may be, as soon as we go back to the origin, we find a few simple basic principles. Two main directions, vertical and horizontal, are the main elements in the shaping of limestone. Therein it distinguishes itself in particular from quartz, but really also from other rocks. Originally, it is deposited in horizontal layers, but is transformed into vertical pillar in the limestone caves. So, more than in the formation of any other mineral, the influence of gravity manifests itself in the formation of limestone.

The Cycle of Limestone in Water

The water that flows down the mountains also takes with it limestone. If the boulders are fairly soft and vulnerable, as can be the case in the Jura Alps (soft limestone mountains in Switzerland), then so much limestone is dissolved that the water continues to be generous with its reserve. It will plaster the bed of the stream with massive layers, which exhibit a play of streaming flow-surfaces, arched here and hollowed out there. In places where the water does not flow, every falling twig will be quickly covered by bizarre white crusts.

However much limestone the water may lose, some residue will always be carried further. We notice this through the formation of a crust in our cooking pots—because our soap-suds form viscous flakes rather than foam. This residual content is also sufficient for fresh water mussels and snails to form their thin shells. In the ocean, it is quite different. We find there those animals which are the great artists of limestone modeling and limestone architecture. Sea water is also so charged with limestone and provides a livelihood for a mass of floating animals, that only here can the full variations of the play of forms come about for which limestone provides the basis.

But every animal species which thoroughly enjoys this modeling in limestone must again pay their tribute to gravity. Even the unicellular foraminifera with their elegant tiny limestone shells—in contrast to the silica-shelled radiolaria, which are very similar in shape—usually live by crawling over the algae and over the bottom of the sea. It is as if limestone pulls all animals down, binds them to the sea bed, and lets them rigidify. There is a multitude of animals living on the sea bed that cover and weigh themselves with increasing loads of limestone. When the animals die, all these shells and small houses sink to the bed of the sea. If they really were to remain there, then they would form massive layers, and the content of chalk in sea water would quickly decrease. In the deep ocean much of the limestone is dissolved through the reaction with carbon dioxide[38] and thus brought again into the cycle. However, we must assume that more limestone is precipitated than dissolved.[39]

We can come now to the following conclusion: The limestone-cycle moves between dissolving and precipitating, be it high in the mountains, or deep in the oceans. In the mountains or in general above sea level, dissolving will dominate; under the ocean, precipitation reigns.

Animal Existence as a Battle with Lime

In the oceans there are animals like the jelly fish, which live floating along with an elegant, transparent body. They keep themselves free from gravity and as it were ignore the lime in the water.

Others, such as mussels and oysters, let themselves be totally surrounded by chalk and bound to the ground. But these, which apparently have been conquered, fight a tough battle. They continuously push the chalk towards the outside and thereby maintain their soft inner bodies.

It is different again for the coral polyps. Although the mussels can mostly creep and even can dig, the coral polyps are more or less prisoner in the great rigidity of limestone, which they deposit around and in themselves. But they are just the ones capable of erecting large structures, in opposition to gravity.

The snails are closely related to the mussels, but in their behavior they achieve a much greater autonomy with respect to lime. Only when life becomes difficult do they retreat into their lime house. In that case, they are motionless and totally surrounded and formed by the lime. However, as soon as the circumstances allow it, they emerge again with most of their bodies and start to glide forth with determination, while carrying their chalk house, as elegant as a piece of armor. By focusing on the group of snails, we can see how far you can come if you are able to maintain the slow gliding forth tenaciously. Many species of snails have left the water and creep all over the land, wherever it is only slightly moist. We find snails in all regions of water: creeping, swimming, floating, and, moreover, distributed across nearly all the land, which means that the snails exceed all other groups of animals in the area of their distribution. This surprising achievement is possible through the peculiar equilibrium this group maintains between autonomy and constraint.

With the squids, the most highly developed relatives of mussels and snails, the liberation from limestone and gravity has been pushed very far and has found a high form of expression. Many species still have a mass of limestone, but this is often very light with a foam-like structure, and is surrounded by their body. We do find certain squids moving along the seabed, but we could just as well call this "walking" on their arms as crawling. What is more, all squids can swim well, and many move graciously and often quickly in any direction through the watery realm.

In the crab group, the question of how to deal with lime is taken up differently. These large species live as full-grown animals on the sea floor, while their larva, in contrast, live as delicate animals, floating in the uppermost layers of the sea. Thus, during their development, these animals slowly sink following gravity. Accordingly, here we also find a sort of coat of armor made of chalk, which is, however, very different from that of the mollusks. Here we do not find a soft body enclosed by an absolutely rigid shell, but a skin that has been hardened, partly through the presence of chalk. But these animals have two ways of escaping from the rigidity of chalk. First, their armor is made so that it assists rather than hinders their movement (well-jointed). Second, the mass of limestone is dissolved before the periodic peeling of the exoskeleton occurs, and is taken in and deposited in the stomach as small stones. After the shedding of their skin, the whole supply of lime is made available to the fresh skin.

With animals such as snails we also find the capability to dissolve the chalk, but here it serves more to make "corrections" to the growth of the shell.

Now, with the vertebrates, we find something totally new. Here an internal skeleton is held together by the spine. The depositing of limestone stands clearly in close relation to the nervous system, as the spine is built around the spinal cord. Moreover, the fact that calcium *phosphate* now becomes prominent must be seen in conjunction. In the lower animals the mineral part of the shell nearly always consists of calcium carbonate. The skeleton of the vertebrates consists mostly of calcium phosphate, and only a small component of calcium carbonate.

If we compare fishes, the first species of animals with an internal skeleton, with the lower animals, we notice a certain drama in the transition to this group. As long as limestone was used to build the shell, in a sense it was focused on the environment. In fishes, however, the limestone is concentrated mainly towards the center of their body. For the greatest part it becomes a cord which goes through the center of the body. From this cord a number of delicate bony processes emerge (ribs). Only in the skull do we find a system of thin skeletal plates which lie on the outside. With this central concentration of limestone, the mobility increases tremendously.

The skeleton has a much greater mechanical function in the higher vertebrates than in the fishes, from the amphibious animals and reptiles upwards. With the limb skeleton of these higher animals, a new relation is made with the environment.

In birds and mammals, the skeleton is brought to its culmination. It is strong and light. Here again we find the typical properties of chalk—solidity, strength—occurring as dead, discarded mass; however, full use is made of these properties, and, therefore, they are overcome. Because of functional connections at the joints, the greatest possible mobility is provided, despite the rigidity of the material. Thanks to the delicate, highly-ordered structure of each separate bone, animals can utilise its firmness, without being burdened with unnecessary substance. The proportions of the whole and the harmonious inter-relation of the muscles and nerves make it possible to use mechanical forces in such a masterful play of motion. We are so used to the sight of a swooping swallow, a squirrel running up a tree trunk, or a deer jumping, that we tend to forget that here we have the expression of an incredible wisdom. And yet no human could make a structure which even remotely compares to them. In the case of chamois and ibex, thus, in the case of those animals which express the impulse of their being with masterly skill in high mountains, the animal kingdom reaches its zenith in mastering the mechanical forces of the earth or gravity. However, this is also a consequence of the surroundings. There is no other realm, where the earth expresses its forces with such bluntness and such vehemence. The animals take up the challenge posed by the earth and not only do they expose themselves to the play of forces but, in a manner of speaking, use this to bring about the ultimate in mobility. Something that is characteristic for all mammals—to use limestone to play with the forces of gravity—is brought to its culmination here.

In the case of the eagle, which surpasses these animals, we seemingly find a further conquering of chalk. In fact, however, something quiet different occurs. Observing the skeleton of such a bird on its own, we see that it is much firmer, denser, or in other words, more earthy than in the case of the vertebrates. Primarily this applies to the substance. However, looking at it from the perspective of the play of forces, the opposite nearly applies. In the case of the typical posture of eagles, i.e., soaring, everything is concentrated on getting the feathers to form the largest possible surface. The animal pulls its feet up, which means nothing less than that it wants to avoid all extension in the direction of gravity. Wings and tail are extended horizontally as much as possible, in a plane which cuts across the direction of gravity. Limestone is not used here to play with the forces of gravity, but to give the feathers a position in which they can radiate horizontally in all directions. Here limestone complies with a play of forces which is directed away from earth and to the periphery of the cosmos. We could also say: here, limestone subjects itself to the forces of silica.

The Diversity of Form in Calcareous Coats, Shells and Skeletons

However heavy the mass of limestone may be with which animals burden themselves and by which they are often bound to the ground, they do, nevertheless, produce the most noble forms in this brittle material. What we see when observing the many lower sea animals is so extremely rich and so full of variation, that we could be led to think that this is an arbitrary game of nature. But careful observation reveals that every form repeats itself time and again with slight variations, and that over the course of time, every form appears again and again. Many snails have apparently arbitrary, wild protuberances, but by these the specialist can unfailingly recognize the species. Even after hundreds of years he would still be able to recognize them without having to relearn anything. This is also the case for such miniature creatures such as the foraminifera, despite the fact that their range of forms is the richest and most enchanting in all of nature. Even in limestone we find shapes that are nearly the same as today's or at least are very related. This gives all these forms a further special meaning.

By developing a great openness, a great devotion, we can experience how seeing becomes hearing. The form of these animals, these snails and mussels, becomes seemingly fluid, and the form becomes a harmony. All these curves become a melodious song; all these repeated turnings become rhythmical sounding harmonies.

We are confronted here with small constructions, consisting of dead material which has been ejected into the outside world by the processes of life, but whose form is "hidden sound"—a harmony trapped in stone, so to speak. If we can now inwardly experience these arches, indentations in the planes, these spirals, then we free the sound from its spell. A world of music then arises in ourselves, which reminds us that our ancestors talked about the harmony of the spheres. Not only did they know that earth and universe formed a noble unity as a huge building, but they also heard its noble order with the delicate, well-balanced relations as an all-pervading music. It is now quite different for us, who have to laboriously attain to this through thinking.

The above gives us a new perspective on the limestone mountains. The rocks, as they are now, are mostly rough, and often present themselves like a wild confusion, but do consist of splintered remains and ruins of forms, which were a direct imprint of the harmony of the spheres.

As in the case of the mussels and coats of lower animals where the most noble forms appear, so it is also with the skeleton of the higher animals and the human being. Every bone displays a magnificent graphic power, but it is also an essential member of a slender construction. On its own, it is more or less a typical unity, but together with a series of other bones it joins into a greater whole. We do not find such flowing lines here as in the case of the mussels and snails. Initially, we are not reminded of music, but much more of the way in which many different sounds come together to form a word, and the words again make a well structured sentence.[40]

Notwithstanding that limestone (chalk) came into being through excretion and consists of dead remains of a vast past, this does not mean that it is unresponsive to formative forces. Here also the well-defined forces of crystallization are at work to create a myriad of glistening surfaces in mute substance. This can reach such an extent that at many places we can cut transparent, purified marble from out of the rocks. After what we have discussed, it cannot come as a surprise that just this material has so often been used by artists to make noble forms, to recreate the delicate curved surfaces of the human body, or to shape the flowing folds of gowns.

Lime in the Human Being

As a young man, Goethe introduced his "Fragments on Physiognomy" (Physiognomischen Fragmente) with a nearly hymn-like glorification of the human form compared to the animal. He describes the human form as a pillar, which is closed off at the top by the cranium, like a dome. He even indicates that this dome must be regarded as mirroring the firmament. He indicates the horizontal direction as being typical for the animal spine. Their head is not carried skillfully balanced, as is with the human being, but is only "attached" or "hangs." Indeed, observing the most upright animals, we will always find departures from the vertical.

Goethe was inspired in his point of view by contemplating the skeletons of humans and animals, thus, by those parts of the organism which contain chalk, forms which have been excreted from the organism. By following Goethe's reflections and extending them, we arrive at some surprising observations. Precisely in our head, which is the part carried highest above the earth and which mirrors the heavens in its shape, we find relatively more lime than in any other part of our body. Moreover, more than at other places, the skeleton here forms an enclosing layer so that a comparison is possible with the lower animals, such as the mussels and sea urchins. Both through the amount of limestone, and by the manner in which it is deposited outwards, the skull shows that it is more related to earth than the rest of our body.[41]

With the legs, however, which are oriented towards the earth and, therefore, are more pillar-like than the rest of the body, it seems as if the typical properties of limestone have been relegated as far as possible into the background. We find limestone here as an internal rod surrounded by soft material. More detailed examination shows it to be a hollow tube. With this shape and with the unique laminated structure at the ends, it shows a principle which also applies in engineering: the greatest possible strength with the least weight. If we were to reform a solid rod out of the material of such a hollow bone, it would turn out to be much thinner and weaker or much shorter. In the wisdom-filled structure of the bone, lime has been arranged in such a way that in every part the capacity to sustain compression and tension has been utilized to the maximum. We can thus say that in the hollow bone of the limbs, limestone has been extended as far as possible without losing strength. For example, a shin bone has been constructed in such a way that it can carry the largest possible weight, notwithstanding it utilizes the least amount of material possible and reaches the longest length possible. If possible, this principle is used to an even further extent in the hip bone. It stands on top of the shin bone like a mountain animal on a narrow summit, and is twice as long as the shin bone. The internal structure of the layers in the region of the hip joint are so impressive that we always find it as the example illustrating the way the human body is built full of wisdom.

When we stand upright, we try to achieve a posture which opposes as strongly as possible that which gravity demands. Thanks to the structure of the skeleton we have described, and the characteristic structure of the bones, we can make use of the relationship of limestone to the earth specifically to conquer the forces of gravity.

Just as around stalactites and stalagmites when we encounter water running from the tip or trickling down inside them, so we find the opposite in the standing human being: an upright standing pillar of liquid, supported by an inner limestone structure. Limestone not only is precipitated from the liquid, but precipitated in such a manner that it can fulfill this supporting function. If the limestone in the skeleton were to remember how it followed its own urge in the stalactites and let itself be utterly governed by gravity, it would wonder at the situation it finds itself in now. Every direction of force is precisely opposite. As seen from the perspective of chalk, the upright human body represents a world wherein everything is up-side-down.

If we bear all this clearly in mind, it is even more surprising to realize how, during sleep, the human being surrenders every resistance to gravity even more radically than do most animals. This is seen in the fact that not only does the human being take on the horizontal position, but lies on its back. Thus, a sleeping person lies in such a position that the form and organization of the skeleton plays as small a role as possible.

What can this mean? The appearance of both these directions signifies for chalk that it surrenders to the forces of gravity. The position of the human being in both these directions also signifies adapting to these forces but not surrendering. The human being accommodates the forces of gravity in form and posture more completely than the animals, only to be able to conquer them more fully.

We have seen how higher animals often show much more finesse playing with the possibilities given them by the skeleton than does the human being. However, the animal does not strive for such a harmonious form or such a balance between heaven and earth. In the case of animals, movement is more important than posture. Every animal specializes in a particular type of movement within a particular environment—the one interacts with the earth, the other works with air, and the third with water.

Because of the complete victory over chalk, the human being is very versatile. The human being is not so much concerned with agile movement, as with bringing together the greatest opposites; to bring harmony to that which could lead to a one-sidedness. Therefore, an imprint of the world lives in the human body—the dome-like human head is an image of the heavens; the pillar-like body expresses the forces of gravity. Up until now, Goethe has rightly described the human being. However, we can find more. The pillar-like structure really only applies to our legs. Consequently, the body there consists primarily of rods. It is remarkable that the chest is somewhat reminiscent of the dome-shaped head but at the same time is constructed from bent, but nevertheless rod-like, bones. We can recognize in the chest a transition form, which is related both to the head and the limbs. In its whole structure a playful element appears.

Nature, too, is not only determined by the contrast heaven—earth. There are many processes, most notably the movement of the sun, wherein a continuous rising and falling takes place, a rhythmic backwards and forwards between the realm of heights and of the depths.

In its form, the chest shows a hardened image of these rhythmical movements. Thus we are confronted with wonderful facts. We have come to know limestone as a substance which ever again makes a fall and is pushed into the realm where only the terrestrial forces are active. But, just this substance is ennobled in the human being until it becomes the carrier of the image of the heavens. In that way, the substance which is continuously being excreted ever again becomes the center of the universe.

What has been discussed so far conforms with age-old traditions: man is a miniature world, and the macrocosm is a vast human. Indeed, everything water does with chalk, e.g. dissolving and precipitation, decomposition and synthesis, our blood does, also, and indeed, following even stricter rules. Even in the shell and crust formation, which occur in so many variations, we find in a very subtle way in human beings. The clotting of blood, through which so many wounds of our skin heal, depends on the cooperation of calcium in our blood.

Once again, we see from this that under all circumstances, a substance not only retains a certain theme, but how living organisms are duty-bound, as it were, to use that very substance which represents a certain needed principle. In the scab, which results from the clotting of blood, only small amounts of calcium may be found. Although calcium as a substance hardly participates, it is an essential assistant in the process of the clotting of blood. It is like it has a "patent" on enclosing with the aid of hard layers. Among the substances it is the principal representative of this process, and creative nature has to take this into account.

During the clotting of blood, blood dies through the influence of a certain contact with the airy environment. It follows the impulses of calcium completely, and it solidifies to an earth-like crust. In contrast, during the formation of the bones blood not only flows while calcium precipitates and remains behind, but in addition blood is renewed right in the center of the rigid world of bones—red blood cells are formed in the bone marrow.

What occurs there? The skeleton has always been rightly seen as sign for the activity of death. However, we must also look in the skeleton for the greatest victory over death. Precisely there where in the midst of the hard bone a space has been created, rejuvenation of blood takes place and we are the strongest manifestation of life in the rigor of death. We also encounter again the "battle with limestone."

The skeleton, which remains a whole much longer than other parts which once were living in the human body, should not be viewed only as a sign of victory of death. In the wonderful play of form, it retains as a memory, an image of the universal human, even of the whole individuality.

Once, there was a people which were so immersed with the divine harmony of their own body, that they could build temples which brought this harmony to expression. No wonder that this people, the ancient Greeks, chose limestone in the form of marble as a building material. And, it really seems to be the case that through building their temples, they expressed the essence of limestone totally. This may be seen from the fact that in these temples the vertical and horizontal direction are strongly emphasized.

However, not only is the essence of limestone expressed here, but also the conquest of chalk, as happens in the human body. The horizontal, flat roof is supported by pillars. It is not the same as what happens in the crust of the earth, where the horizontal layers such as the stalactites are formed through the influence of gravity. Rather, it is related to the human form where the pillar-form is directed to conquering gravity and the most earthy part is carried high above the earth.

Chemistry of Calcium

Lime as an Alkaline Substance

When we direct our attention to the chemistry of the crust of the earth, then again we must naturally pay attention first to reactions of an acidic or alkaline nature.

Just as the most colossal transformations of the earth's crust may be the result of small events, which, because of endless repetition have far-reaching effects, so chemical activity in the minerals has a considerable effect in time, even if the activity may be tiny.

The more silica the soil contains, the more acidic it will tend to react. We find this in regions with a high sand content in the soil. There, the tendency exists of forming dead humus, as is the case with a high heather moor or peat bog. Because of the acid environment, processes tend to come to a standstill, and as a result humus tends to become conserved in such soils. The more limestone in the soil the more alkaline it will react. Despite the fact that limestone is a salt [usually considered neutral], its alkaline aspect dominates the acid. Such soils activate the process of decomposition, so that a humus deficiency easily occurs.

In accordance with its alkaline nature, limestone has the property of being "greedy." It binds all sorts of substances, especially those which react as acids. The most important example is carbon dioxide gas.

By heating lime, its alkaline nature and its tendency to combine with other substances greatly increases. We will discuss later on how quicklime (calcium oxide) has a very strong burning effect.

Calcium as Element

With chemical analysis, all lime-containing substances show the presence of the element calcium, one of the so called light metals. By expending quite a lot of energy we can extract this metal from molten calcium chloride ($CaCl_2$) in electrolysis. It then exists as brittle, small metallic pieces. If we allow these pieces of metal to lie exposed to moist air, they do not liquefy as does elemental sodium metal. They slowly become dull and turn into a pure white, earthy mass. Eventually, they can even fall apart as a white powder.

The appearance of the color white with the calcium compounds is so conspicuous that we must devote a special discussion to it.

The processes which have taken place are mainly: oxidation (to CaO), formation of hydroxide [hydrated oxide, $Ca(OH)_2$] and formation of limestone [calcium carbonate, $CaCO_3$].

We can ignite elemental calcium when we direct the flame of an oxy-hydrogen burner onto a small mass of calcium. We then observe a very sudden, very violent reddish glow. If we are lucky, we can even briefly see a series of red sparks erupt above the whole. Afterwards, a dull pure white mass of calcium oxide remains.

A piece of calcium put into water sinks. It immediately starts to evolve hydrogen gas, just like sodium and potassium, but not as violently. Insofar as the metal disappears and the water becomes basic, it resembles the alkali metals. But, from another perspective, through calcium water acquires totally different properties than through sodium. If we added sufficient metal, then the water becomes turbid and white flakes appear, which sink to the bottom. If we observe the result of the experiment the next day, we notice that the water is again clear and that the bottom of the beaker is covered with a white mass of calcium hydroxide, while the surface is covered with a cracked, thin crust of limestone. The liquid thus seals itself off with a layer of calcium carbonate. Through lime, water acquires earthy, or earth-seeking properties. Furthermore, from this experiment it follows that calcium can reduce the amount of water; only part of the formed hydroxide remains in solution.

This experiment again demonstrates that calcium will only be at peace when it has transformed into a carbonate.

Again, we see here an example of our observation that the chemical elements in our elemental circle (see Chaps. VII.3 and VII.4) present us with images of the realms of nature from which they come. We will see further that in our experiments it is even possible to present phenomena like the formation of stalactites.

Calcinated Lime and Slaking of Lime

If we intensely heat a piece of limestone, a shell, or some marble chunks with the aid of a powerful burner, then we will soon observe an intense white glow. After a short while, the stone will appear altered where we heated it. Now, although it seems to retain its form, it is easily crumbled into dust. Some carbon dioxide has escaped, and what remains is "calcinated" (roasted) lime (calcium oxide).

Instead of using self-made burnt lime, it is sometimes more convenient to buy it, because you can work with larger amounts (although you can roast it dramatically yourself, in a home-built oven).

After having let the burnt lime cool, we pour some water over it; this water will be quickly soaked up, while the resulting mass swells and crumbles. Surprisingly, what remains looks quite dry. All this happens accompanied by massive evolution of steam (water vapor), heating, and fizzing and sizzling. Here we have the so called "slaking" of lime, the change from "roasted" lime (calcium oxide) into "slaked" lime (calcium hydroxide).

As a strongly alkaline (basic) substance, the roasted lime has an all-consuming, burning, destructive effect on organic substances. In times of war or revolutions, the many corpses were laid in lime-pits to be quickly destroyed. Calcinated (roasted) lime is also used in tanneries to remove hair (depilate) from the skins.

A more delicate use of burnt lime is in compost heaps. The remains of the vegetable substances rot away more evenly and quickly.

Calcium hydroxide and Carbon dioxide

Slaked lime, shaken with water, results in a white, opaque suspension which we call slaked lime. The clear liquid which results as the opaque mass settles out, is called "limewater." This is a very dilute solution of slaked lime.

If we pour some limewater into a conical flask and blow exhaled breath into the upper part and stopper and shake the flask, then the liquid becomes turbid—limestone (calcium carbonate) has been formed in a reaction with the carbon dioxide in our breath. We use this experiment to demonstrate the presence of carbon dioxide.

However, if we *continue* to blow into the turbid solution, it becomes clear again! The calcium carbonate has changed by further reaction with the carbon dioxide, into calcium bicarbonate, which is quite soluble in water.

The whole experiment is quicker and clearer if we use carbon dioxide from a gas cylinder, and let this bubble through the limewater.

As we only know of calcium bicarbonate as a solution, we may also say: the slow, insoluble calcium carbonate is made mobile and is carried away by the addition of carbon dioxide and water. In the beginning of this chapter we already discussed the important role this process plays in nature.

When the excess of carbon dioxide disappears from the solution simply through evaporation or heating, then the calcium carbonate precipitates as limestone. This provides the basis for the formation of stalactites/stalagmites and a lime crust on pots and teakettles in areas with "hard" water.

In the living soil this process of absorbing and releasing water also plays a part.

A Few Illustrative Experiments

If we drop a piece of marble into a solution of hydrochloric acid of moderate strength (6N), it will dissolve with a strong foaming and bubbling effervescence.

calcium carbonate + hydrochloric acid → calcium chloride + carbon dioxide

If instead of marble, we use a number of shells, then we observe a large amount of froth being formed. If we had used a tall glass cylinder, we would have seen the froth move up and spill over the edge in a form similar to a snail-like animal.

The calcium chloride thus formed is a very soluble salt. We can, therefore, use it as starting material for a series of very characteristic experiments.

For example, we can pour a fairly concentrated solution of calcium chloride into a glass (hydrometer) cylinder, place this over an illuminating apparatus, and then add some concentrated sodium hydroxide solution. We then get a surprising quantity of a thick, very white precipitate of calcium hydroxide. In this way the tendency of calcium to precipitate can be nicely demonstrated.

We can then make a similar set-up and pour a layer of dilute ammonium sulfate solution onto a concentrated solution of calcium chloride. We then observe a particularly beautiful precipitate which forms at the boundary. Quite slowly very white, small, delicate flakes of gypsum (calcium sulfate) are formed. The silent falling of the flake-like crystals, the formation of white, loose layers, reminds one strongly of the phenomena occurring when it snows.

A very interesting, but somewhat violent experiment is the following: We use a fairly large glass beaker and fill it not more than one-quarter full with a concentrated calcium chloride solution. Then we slowly pour concentrated sulfuric acid along the wall of the cylinder. A very thick mass of gypsum is formed, which appears to us as a romantic landscape of stalactites and stalagmites. We must however work in the fume hood, because of the large amounts of hydrochloric acid gas being evolved. Moreover, we must take care that the glass does not break because of the strong heating. Afterwards, it is usually possible to carefully pour off the water, without the precipitate falling to pieces.

There are probably better methods by which you can imitate the formation of stalactites. For example, we might hang a piece of calcium chloride in a sodium carbonate solution or in a sodium hydroxide solution, but so far I have not found a good way to do this.

At any rate, it is possible through experiments to rebuild the landscapes that belong to the various calcium compounds.

In nature the chalk processes continuously play between earth and air. Thus, a part of the calcium compounds are quite soluble, and a small group only fairly soluble.

quite soluble	fairly soluble	hard to dissolve
calcium chloride	calcium hydroxide	calcium carbonate
calcium nitrate	calcium sulfate	calcium orthophosphate
calcium bicarbonate		

Something about the Uses of Calcareous Substances

It is self evident that the interchange to and fro between water and earth constitutes the main theme of lime technology. Both in the use of mortar, and of cement and gypsum, a watery paste is made which after a time congeals to a brick-hard mass. In the whole context it is rather surprising that in all three cases activation occurs through pre-treatment with water.

The processes during the production and use of mortar have already been discussed.[42] The slow hardening is the result of the absorption of carbon dioxide and the formation of calcium carbonate. We see here the formation of a salt from an acid and a base.

Cement is prepared by roasting a mixture of lime and clay and then grinding it to a powder. In use, this is mixed with water to a paste; then during hardening various silicates are formed. We could call the most common application of lime and cement in construction a making of artificial mountains and caves.

Plaster of Paris is made by heating gypsum (calcium sulfate crystals) for such a length of time that part of the water of crystallization is driven off. In the process, a powder is formed. By adding water to this powder the water of crystallization is restored, while the mass hardens.

Plaster is used for many purposes, especially, however, to aid or even to replace the skeleton with its solidity (after a broken bone).

In any case in the technology of lime we find a beautiful example of how the course of quite different processes can be led and determined by the same motive. This motive is connected to the character of lime. The variety of the processes depends on the interaction of various substances and forces in nature.

The discovery of reinforced concrete may be seen as the culmination of lime technology. The solid mass of concrete can really only be used to stack masses upon one another; it can withstand compression, but not tension. Reinforced concrete owes its capability to withstand tension to the embedded iron (steel) rods. Therefore, it can be used for construction of a large span such as bridges, halls of stations, etc., constructions that could never be built using stone. In reinforced concrete, we have found a substance that will do justice to the highest mechanical demands.

In reinforced concrete we have a substance that is composed of the most abundant components of the crust of the earth: lime, aluminum silicate, water, iron, and silicic acid in the form of sand.

In building large constructions like bridges, we are doing something similar to the animal organism when it builds a skeleton, which makes standing and agile movement possible. It is possible to maintain that in the modern constructions using reinforced concrete. Uprightness has reached technical perfection.

Through further reflection we also find the jump realized in engineering. To that end we must view a jump for what it really is: namely bridging the gap between two distant points without any support in between. In engineering we find a sort of image of jumping in the construction of bridges. We may really consider bridges as congealed jumps.

In human beings and animals, lime first had to become rigid to make possible standing as well as a remarkable mobility. Similarly, construction in reinforced concrete had to be developed to enable standing and jumping, in the above sense, to unfold in such a majestic way.

A surprising development has taken place in the lime industry. Initially, lime mortar was used to connect blocks of different materials which were piled one on top of another. For the constructive elements, wood or iron were used. Nowadays the material to be used is mixed with lime to a compound and even interwoven with iron. Only now, those things which come about in nature with the help of lime are being realized in engineering—we can think of the enclosing of space, as with mussels, the heaping of matter upon itself, as with coral, and mechanical construction, as in the case of skeletons.

The essence of lime has only been realized in engineering through the discovery of concrete. Now also in engineering do we find lime involved in the impressive interaction with gravity and its conquest.

Lime and Light—Contrast to Silicon

When we let calcium metal lie in the open or roast sea shells or marble, or when various calcium compounds precipitate from solutions, the white color is very obvious.

We use chalks to write white on black and to white-wash walls.

Also, in the case of lime and marble, although other colors also occur, we often find a conspicuous bright, even white color. The white color emerges especially brilliantly when its under-surface is strongly illuminated, e.g., a chalk cliff face in the afternoon sun. The eye is blinded by the dazzling play of light, while heat is hardly noticeable as would be the case with a dark stone. The white surface throws light back with force and what is more, there is hardly any change in the layer directly underneath.

If we strongly heat the ends of a piece of real chalk with a powerful flame, it will flare with a dazzling whiteness. Initially, calcined chalk is formed, and that causes the strong flaring. It is an incandescence without transitions; we only see a white shining.

We have previously established that in the case of lime, where light is hardly converted to heat. We now see that heat can be converted to light. (This phenomena was formerly used under the name of Drummond's chalk.)

From the beginning it has been obvious how lime falls, submits to gravity, is condensed. It is really self evident, when we know the contrast between gravity and levity, that a substance with such a connection to gravity should ward off light. Lime achieves this through its white color. While quartz combines a great degree of transparency to light with the tendency to negate or conquer gravity, in lime we find precisely the opposite: warding off light combined with the tendency to submit to gravity.

Both substances exhibit in the characteristic way in which they appear, as it were, the contrast of gravity and levity, but each in an opposite manner.

Lime as Opposed to Hydrogen

Something surprising emerges if we consider what goes on when a corpse decomposes. To a large extent the substance volatilizes as a gas or as vapor, while the skeleton remains for a long time. Hydrogen is to a great extent responsible that the softer parts disintegrate and become completely formless. In contrast, lime serves to retain the form of the mineral part.

Here we encounter most clearly the contrast between hydrogen and lime, which are situated opposite each other in our elemental circle. Lime forms boundaries, tends to condense, generates rigidity, goes for the earth. Hydrogen breaches boundaries, scatters, flees the earth, sets everything in motion.

Phosphorus and Magnesium

On the Availability and Circulation of Phosphorus in Nature

We find phosphorus in nature as hard rock which is difficult to mine. In the granitic basement rock, we find it as apatite. This name indicates a few compounds, which are only rarely found in great masses in specific places, but are rather common in many places in small amounts. Their composition is $Ca_5(PO_4)_3Cl$ or $Ca_5(PO_4)_3F$.

Through weathering of the rock, calcium phosphate reaches the soils of woods and arable land.

For the extraction of phosphorus compounds, phosphorite (apatite)[43] is of particular importance. It often appears as a nodule of stone and has a variable amount of calcium phosphate (up to 80%). The bones of the vertebrates consist for the most part of calcium phosphate (predominantly hydroxyapatite).

The guano from the Galapagos Islands, consisting of excrement and the remains of sea birds, is particularly rich in phosphorus because of the calcium phosphate of the fish skeletons. Over time, the mass of guano is becoming more and more mineralized. While the more reactive parts of the excrements can disperse because of decomposition, the percentage of calcium phosphate increases continuously, so that in the end a phosphorus-like mineral comes about.

We must surmise that a lot of phosphorus also sinks to the bottom of the sea with the skeletons of the fish.

Such facts indicate—and later on it will again become apparent—that phosphorus has an intimate relation with the earth element and, therefore, even has the tendency to disengage itself from the processes of life.

We can quite understand that such a substance is very suitable to engage itself with lime.

Even when phosphorus participates in life processes and assists, for example, with the production of certain proteins, it still maintain its typical inorganic form to a high degree. While sulfur has the tendency to go through various forms and, consequently, experiences many bondings and dissolutions, phosphorus will predominantly act as phosphoric acid or phosphate. The same applies to arable soil, where phosphate is found mostly in a very insoluble form.

In living beings we find phosphorus throughout the whole body. In plants most of the phosphorus is found in the nucleus of the cell and in the seeds. In humans we find phosphorus particularly in the head. Partly this is due to the predominance of the skeleton but also due to the concentration of nerves. In nerve tissue we find slightly more phosphorus than in other tissues.

In general we find phosphorus at those places in the organism where we observe a certain concentrating effect, a hardening and a coming to rest.

What happens when we free phosphorus from its stiffness and bring it to its elemental state?

Phosphorus as Element—Fire and Light Phenomena

To prepare elemental phosphorus we need an electric oven. In a small space we create the very high temperature that is necessary to bring about an acid displacement with the aid of silicic acid and a reduction with the aid of charcoal. As starting material, we can use bonemeal and mix this with sand and charcoal. Introducing this mixture in the oven releases a gas which is a mixture of (gaseous) phosphorus and carbon monoxide, while molten calcium silicate flows away.

It is always important to be aware of the dramatic changes that can takes place during such a process. We start with a very rigid, immobile, earth-like substance and end up with a pale, waxy material. The white phosphorus which results is so unstable that we can only keep it under water and in the dark, a substance that appears as something unreal, without essence.

Red phosphorus, into which white phosphorus slowly changes under influence of light and warmth, is quite different. We can compare red phosphorus, as far as its properties are concerned, with carbon and sulfur. It is very combustible and fairly stable at room temperature.

Phosphorus in both forms clearly distinguishes itself from both carbon and sulfur, in that it is never found in nature in elemental form.

Red phosphorus is difficult to melt and not very soluble. It evaporates only at fairly high temperatures. If we force the evaporation and lead the vapor through cold water, we get white phosphorus again. It melts already at 44°C and readily dissolves in carbon disulfide, and begins to vaporize at room temperature.

If we heat red phosphorus exposed to air, it will ignite at around 260°C (only at about 435°C, if very pure). It has a bright, yellow-white flame, while simultaneously a dense white smoke snakes upwards. This smoke also distinguishes the flame of phosphorus fundamentally from those of carbon and sulfur.

It is hardly necessary to ignite white phosphorus–it smokes immediately. It will even ignite with a slight rubbing or a little heat. The flash point is quite low, around 60°C. We must be aware of this when handling phosphorus. It is best to hold the small white stick with a wet cloth. If you use a pair of tongs, then the stick may jump away and ignite due to friction. If we use our fingers, then we expose ourselves to a DOUBLE DANGER: In the first place one can get burns, and in the second place one may be poisoned. White phosphorus should be cut under water. We can, of course, ignite white phosphorus with a flame. The same flame is seen as in the case of red phosphorus — only everything is much more violent. This fire is very hot.

If we collect the white smoke, then we get a white powder-like substance, phosphorus pentoxide (P_2O_5). This substance vaporizes directly at 250°C, skipping the liquid state. Burning white phosphorus thus melts more easily and is more volatile than its oxide (unusual for oxides).

We can understand what is expressed herein better by recalling what has previously been said about combustion and rusting.[44] In combustion, other than in the case of rusting, substances are dispersed in all directions. In part that is also the case with phosphorus. While the oxides of carbon and sulfur are gaseous and thus continue the dispersion after combustion, the oxide of phosphorus is more solid and more earth-like than white phosphorus itself. Sulfur and carbon move two stages higher on the ladder of the four elements: from earth to air. In the case of white phosphorus, combustion begins with an ascent which is immediately followed by a still deeper fall back into the earth element. While sulfur is carried along by the mobility of oxygen in combustion, in the case of phosphorus, oxygen is congealed to an earthy substance, like the case of a metal rusting. Here we have the surprising phenomena of a combustion headed towards gravity and seemingly proceeding as a fall. It is always possible to enhance the peculiarity of such phenomena by choosing a special experimental set-up. For example, we can place a cool glass bell-jar over the flame. First, we see the smoke rise and then trickle down the sides. We can observe the phenomenon of the downwards flowing of smoke even more simply, by holding a piece of white phosphorus in the air. We can make the phenomenon even more striking by writing carefully on a vertical black surface with a piece of white phosphorus. Immediately, we observe delicate wisps of smoke sinking downward.

That in itself is surprising enough. If we observe white phosphorus in the dark, then we encounter the extraordinary phenomenon that it is shining with a weak light. If we strike a hard surface with white phosphorus, then we observe luminous lines. If we then blow along the luminous lines, then we observe waves of light and dark following each other. We have here a slow oxidation without any significant increase in temperature. We could speak of cold light. Phosphorus slowly evaporates, and accompanied by light phenomena, changes to a smoke-like oxide.

It is possible to demonstrate this phenomenon of quiet glowing in a very impressive way, if we modify the following experiment,[45] which is normally conducted as follows: white phosphorus is dissolved in carbon disulfide to give a concentrated solution. Some of this solution is then poured onto a piece of filter paper which is waved about in air to dry it quickly. After a while we again observe white smoke descending and immediately afterwards the phosphorus ignites and the paper burns, too [CAUTION]. If we dilute the solution, however, then combustion can occur without the paper burning. In this case a very remarkable painting of yellow and brown patches remains behind on the paper as a sort of moonscape. If we dilute further, the occurrence of flames can be completely avoided. We will then observe a strong phosphorescence, at least in the dark. We are reminded of the moon and how it shines at night through a fog.

If we spill some of the solution in our "demonstration box" this will look, after a while, like a glimmering enchanted castle.

After the experiments, as a safety measure, we must remove the phosphorus residues, burn the used paper in a fume hood, and play the flame of the Bunsen burner over the work surface.

In such experiments it is especially important that the glowing stops after a certain time, because all the phosphorus has disappeared. Thus, the white phosphorus volatilizes, producing cold light and smoke.

With phosphorus, it is very unexpected that light and fire phenomena often occur in conjunction with water. For example, it is even possible to ignite white phosphorus with the aid of hot water. Place a piece of phosphorus on a stone which water can easily run off, and then pour some boiling water over it. As soon as the water has run off, small white flames and smoke appear. If we put a piece of phosphorus in a test tube with water and blow some air through it, then we can see it burn even under water. We can also put some potassium chlorate in a flask with cold water and then add some phosphorus [VERY DANGEROUS]. Then we add some concentrated sulfuric acid with the aid of a pipette to the potassium chlorate. Through the formation of chlorine, phosphorus is oxidized accompanied by glowing.

Images of the Night Sky during Experiments with Phosphorus

New phenomena occur in a variant of another well-known experiment, the self-ignition of hydrogen phosphide. For this experiment we need some water, sodium hydroxide, and a piece of white phosphorus. These are heated together until the water boils. Hydrogen phosphide gas (PH_3) is formed.

Mostly, the experiment is done in such a way that the gas is first led through cold water. With each small bubble, we see a small flash of flame and a smoke ring. Now it is possible to do this experiment in a large, tall-form beaker which is quite full. If we then darken the room, we see an amazing display of fireworks, a violent flurry of sparks. We are not sure if we should compare this to miniature lighting or dancing stars. If we increase the scale of the experiment, then we will observe small, unexpectedly sparkling stars in the vapor up to a height of about 10 cm above the liquid. If we turn off the heat after a time, everything calms down to rest. Then, if it is dark enough, we can observe for quite a while, small, delicate glowing points on the beaker wall. If the conditions are appropriate, then we can get the impression of being confronted with a miniature starry heavens. If there are also small moving glowing points in the water, then we are strongly reminded of marine phosphorescence. Those of us who have experienced this from close up know that this reminds us surprisingly of a clear night sky full of stars. We find this often in descriptions of the open ocean—a dark space filled with glowing points, which reminds us of the starry heavens. This after-glow of star-like points is such an intimate phenomena that we can only clearly observe it from close by. However, it is also possible to create an image of the starry heavens which can be seen very clearly and from a great distance. We again start with a dilute solution of phosphorus in carbon disulfide. We must pour this solution from some height or must sprinkle it, so that the drops spatter about. After they have dried, they leave glowing points behind. It is even possible that we have spattered our own clothes [NOT RECOMMENDED]; then we would appear like a magician in a mantel of stars. Perhaps this experiment is not as elegant or clear as the others and, therefore, less convincing. However, we must take into account that here we are also dealing with a "disposition" of phosphorus. There is no other substance with which we can do anything similar.

Properties of Magnesium as Element—Phenomena and Experiments

A phenomena just as surprising as the quiet glowing of phosphorus is the combustion of magnesium. This silent brilliance is the result of oxidation at normal temperatures, something we especially encounter in metals. But that it can be ignited and burn with a flame usually belongs to the properties of non-metals. It is, therefore, only consistent that magnesium hardly rusts and that it is only slightly attacked by the atmosphere, in contrast to the closely related calcium where this occurs to a high degree. If we sprinkle some magnesium powder in a gas flame, then we observe it ignite as dazzling, short, straight streaks. A larger amount of powder produces a blinding white light while white smoke shoots into the air. This light is so brilliant that it can be dangerous to the eyes. It is very characteristic of magnesium that we can produce light pulses. There are absolutely no transitions. Suddenly, light is there, in all its intensity and brilliance, and it is extinguished just as quickly.

Light from magnesium is distinguished by a strong chemical effect, so that it can even be used in photography.

If we hold the end of a strip of magnesium in a flame, then it will start to burn with a brilliant, but *quiet* flame. The flame eats its way along the strip and leaves behind a white ash, which still weakly holds its form. Above the flame, a white column of smoke goes up. Both smoke and ash are magnesium oxide (MgO). As with soot, we can deposit this "smoke" on a surface; it gets covered with a layer that surpasses all other substances in whiteness, just as soot outdoes all other substances in blackness. We can achieve a very splendid effect, if we first blacken a piece of glass with soot from a smoky flame, and then make part of the rod white with magnesium smoke. A special impression is presented when we compare two well-illuminated beakers, one of which is blackened on the inside by soot, the other whitened by magnesium smoke.

We have to mention that the magnesium flame reaches a high temperature but in comparison only radiates a little heat. However, magnesium produces the strongest intensity of light of all possible flames.

The whitest substance possible is produced by the most brilliant flame. The substance that reflects light most strongly comes about in a process which radiates the most light. Connected with the reflection of light is the fact that magnesium oxide, more than other substances, remains cool when illuminated. In contrast, soot absorbs more light than any other substance and as a consequence, becomes hotter through illumination.

One of the surprising facts in conjunction with the light of the magnesium flame, is that it strongly resembles sunlight in color, intensity, and the nature of its effects. It is possible to create black shadows in sunlight, if we use enough magnesium.

In his textbook on inorganic chemistry Hoffman[46] writes: "A disk made from burning magnesium, which resembles the apparent size of the sun as seen from sea level, has the same effect at this point as the sun at 9°35' above the horizon on a cloudless day."

In this chapter, we have already previously established that phosphorus is a substance which presents an image of the night sky. Now we see that magnesium may be used to imitate the sun.

The Occurrence of Magnesium Compounds in Earth and in Living Organisms—Contrast to Lime

Magnesium has the same position in the periodic system as has calcium. And, in fact, there exists not only a correspondence in chemical properties, but in rocks we also often find magnesium and calcium compounds together. The mineral most well-known for this is dolomite. The most typical magnesium minerals are quite different from those of calcium. A very curious mineral is asbestos, a magnesium silicate. From it we can extract delicate, glittering fibers, which are so flexible and tough that they can be spun. Here we have a mineral with properties which are completely opposite to those of other minerals. These are mostly characterized by a certain hardness and brittleness, so that small pieces easily break off. In general they resist changes of form. Asbestos fibers, however, are quite amenable to changes in form, without their losing their inner coherence. Other important properties are its resistance to fire and chemical influences.[47] Some magnesium silicates are also very remarkable because of their leather or wood-like appearance.

Magnesium performs its most important task in chlorophyll. From a chemical perspective, chlorophyll and hemoglobin are very similar. While iron forms the center of its chemical structure in the case of hemoglobin, magnesium performs this function in chlorophyll.

We can only understand the structure and life of the plants well when we see them as gestures of the earth. Through the plants, the earth opens itself to the sun. The plant opens to the cosmos not only with respect to form, but in chlorophyll it is also carrier of a substance which is completely receptive to light. Through the green of the plants, the earth has formed a relationship with the sun. The chemistry of the earth has become fructified by the sun.

We have arrived at a point where it becomes clearly visible, that not only is there a certain relationship between calcium and magnesium, but also a great contrast. Calcium takes quite a different direction in the human body than does magnesium in the plant. That becomes clear when we look at the skeleton. In the head region, calcium creates more and more enclosing, protective, forms; towards the feet, it accommodates gravity. Calcium in the limbs assists the human body just as much with its interaction with gravity as the magnesium compounds do with the interaction with light in the plant.

The contrast between calcium and magnesium compounds is very marked in the oceans. In the animals living at the bottom of the sea, calcium comes into its own to the highest degree, while in the chlorophyll of the plankton algae just under the surface, we find magnesium doing its work. Through the lime, the animals living at the bottom of the sea are not only bound to the earth but have turned away from the light as much as possible. In contrast, the algae live and float in the illuminated part of the oceans and are, therefore, free of the influence of gravity.

If magnesium acts in its characteristic manner it is always found to enhance the light processes and oppose the forces of gravity. It is as if it would like to leave the earth and follow the sun. It helps the plants form a connection between earth substance and the activity of the sun.

From a different viewpoint, we can now understand why in the case of typical land animals living in full sunlight the skeleton, for the most part, is found within the body. Here also lime recedes from the light. This occurs especially in the case of the eye-sockets.

We could contrast this with the teeth, by which the skeleton breaks through the skin. It is precisely here that magnesium plays an important role alongside calcium.

After all this we must state that the name "earth alkali" for calcium has been well-chosen, but that magnesium should really have been called "light/photo alkali."

We can now also better understand the structure of asbestos. In nature the general rule holds: the substance of plants or animals whose lifestyle is totally devoted to light takes on a linear form or even becomes fibrous. Clear examples are seed and fruit fibers in the case of plants, the very delicate protuberances of plankton animals, etc. In asbestos we have a substance which of its own accord has taken on a structure as if it were formed by light. The asbestos fibers are just as much an expression of the tendency of magnesium to be open to the influences of light, as the stalactite is of the disposition of lime to follow the rule of gravity. This disposition of magnesium is further strengthened by silica. We can understand the fibrous structure very clearly by recalling the great contrast we have between light and gravity. In the seed fibers of a dandelion we find a minimum of substance, but due to its fine structure, it stretches itself quite far out into space. We can hardly speak of gravity here, while furthermore it appears as if in its radial structure these fibers follow the gesture of light. Moreover, such a spherical seed-head is very conspicuous. Such structures offer a lot for our eyes, although not so much to our sense of touch.

Phosphorus as Creator of an Equilibrium between Heaven and Earth and Herald of the Conscious Spirit

We have seen that phosphorus mostly occurs together with calcium. That is so for its appearance in geology and in living beings. Phosphorus and chalk form an insoluble salt. This is an expression of the tendency of both substances to withdraw from the flowing, living cycle and to harden, stiffen into the earth element.

We can recognize the same tendency in the formation of smoke when phosphorus burns. The phosphorus flame starts off by radiating and volatalizing, which is generally characteristic for fire, but almost immediately a hardening sets in, a sort of fall into the abyss. In this blazing up and immediately being repelled, the nature of phosphorus becomes clearly manifest. It is particularly telling that the tendency to rigidify extends into the state of fire.

From an alchemical point of view, we can view the phosphorus flame as a sulfur-process, into which a salt-process has penetrated.

The phosphorus pentoxide which forms during burning is especially hygroscopic. With water, the oxide forms various phosphorus acids which are all solids at normal temperatures. Through phosphorus pentoxide, water is also brought into the sphere of the earth element.

We can now understand why most phosphates are insoluble, and cannot free themselves from the earth element.

If we now want to consider more fully the task which phosphorus fulfills in living organisms, then we must first recognize how phosphorus keeps its affinity to light while becoming encapsulated in the earth element. Important facts which can serve as a basis have already been discussed in conjunction with calcium. We had to give quite a lot of attention to the amazing structure of the bones, which give our body the possibility to use the forces of the earth. The lime-containing substances are used to withstand the forces of gravity to which they are in fact related. In bones we find two earthly properties of chalk harmoniously brought into balance: First, the tendency to solidify from the liquid state to the solid, and thereby to become open to the influences of gravity; and second, to vigorously retain certain structures and thereby be able to withstand gravity. Only then does chalk become useful to the motor system of animals; through this, the vertebrates have achieved their mastery of control over the mechanical forces, and the birds have achieved their ability to take off from the earth.

This principle reaches its fullest harmony in the human being. Although it is governed here by the laws of gravity more than in animals, the human celebrates its greatest triumph over gravity in an upright posture. And that also demonstrates a wonderful equilibrium, in that the human being can just as well orient itself towards levity, as to gravity. The human skeleton is not only built so that the whole form becomes the bearer of the brain as organ of consciousness and thinking, but as a sense organ for light, the eye is also carried above all other organs and parts of the body.

We may see in this phenomenon the highest expression of the nature of phosphorus. Preferably calcium phosphate ($Ca_3(PO_4)_2$) is used to accomplish such tasks. Where calcium carbonate ($CaCO_3$) predominates, the influence of gravity is more in evidence. Thus, it becomes clear that phosphorus appears as the agent of light or levity, even in the most hardened, earth-like part of our organism. It represents, as it were, the interests of light in the earth element.

In the same way that phosphorus remains true to light in the sphere of the earth element, it contributes to the ability of the human organism to have consciousness and thinking, because an upright posture is also necessary for this. Phosphorus, thus, serves both light and thinking, a connection that must not in any way be seen as being accidental.[48]

While phosphorus is more or less immobilized in the skeleton, in the nervous system and especially in the brain, its compounds are continuously participating in processes. These processes are again intimately related to us becoming conscious of perceptions and developing thoughts. Phosphorus is so profoundly connected with our waking consciousness, that homeopathy prescribes certain phosphorus remedies to stimulate proper waking up.

In general there is a contradiction between the occurrence of material processes and the unfolding of consciousness: with our consciousness (nor in particular, our thinking consciousness), we do not participate in the metabolic processes taking place in our body. And yet the material structure of our body presents itself as carrier of our consciousness, and even as carrier of our spirit. We could regard that as a miracle. In its body, the human spirit has not only developed a tool with which it develops a perceptive and understanding consciousness, but also one it can use to realize its own ideas. Wherever the body serves as a tool of the conscious spirit in this way, the path is prepared with the help of phosphorus. This becomes obvious by realizing that with every muscle activity, the accompanying material processes begin with changes of phosphorus compounds. Phosphorus, thus, is a substance that not only helps us to gain insight into the ideas of things, but even helps us to realize these ideas. Phosphorus is, as it were, the material substratum for our ability to take initiatives.

What happens when an idea lights up in our mind? If this idea relates to a natural process or to beings of nature, then we can speak of the creative power of nature becoming conscious, which otherwise acts unconsciously. When we become aware of the content of the creative power of nature, it is then freed from its relationship with nature, stripped of any effect. The force of the creative reality, in our consciousness fades away to an image. The ability to bring this about lies in the remarkable constitution of our head[49], which is polar to that of the abdomen. During protein synthesis which takes place in the metabolism, inorganic substances are broken down and the substances thus become accessible to the higher laws of life; in the head region, a falling back of the substance into the inorganic takes place. During metabolism in the sphere where sulfur plays a leading role, substance is split off from the realm of the inorganic and is brought into the realm of life so that the processes of life act upon it; in the head region, life processes are being more or less divorced from substances. These forces are set free and become available to our capacity to think.

By virtue of this severing of our own constructive forces from being subject to matter, it is possible to free the ideas from nature and use them freely.

If we then proceed to put these ideas into practice, we are doing consciously what occurs in nature unconsciously. A person who has trained himself in dealing with ideas in a sovereign way, takes on the task of continuing and finishing creation. He does not remain a small cog in the large mechanism of nature, as the other creatures generally are, but rather he places himself at the beginning of a whole new development of events.

If we summarize all the insights and phenomena discussed thus far, then it becomes obvious that the material properties of phosphorus already indicate its task to a high degree. As with the occurrence of light in the case of white phosphorus, which is accompanied by a hardening and "putting down" of the substance, so also hardening and putting down of substance in our heads are prerequisites for ideas lighting up in our consciousness.

The Magnesium Flame with Regard to Photosynthesis

Magnesium as Representative of the Sun

There is a close connection between the properties of the magnesium flame and the task of magnesium in photosynthesis. Here again we see a very significant example of a process which follows one of the main rules of chemistry. In general when we observe a chemical element, that substance will have been extracted from that part of nature to which it belongs. By inducing the element back into the larger context of nature, e.g., through oxidation, then an image appears of that part of nature from which it originally came. Sodium surrounds itself with a mass of water, calcium forms earthy, white crusts. We can now take up magnesium by saying: With magnesium it is easier than with any other substance to imitate sunlight. The most important task of magnesium in chlorophyll is to prepare the earth to absorb and assimilate sunlight. Burning magnesium must "of necessity" bring forth the image of the sun.

As always we find in technology the realization of such insights, without engineers being consciously aware of them. When we want to illuminate a large tract of land for military purposes and we launch a star-shell, an artificial sun, we use magnesium.

We have previously seen that, if we direct a stream of oxygen gas onto a glowing piece of charcoal, we can also create a clear image of the sun. The chemical elements which are directly involved in photosynthesis—oxygen, carbon and magnesium—are also the elements, which are best suited to produce an image of the sun.

While in the case of magnesium we have sudden transitions, fierce flarings-up, in the case of burning charcoal we find very gradual transitions. Often carbon does not become so hot as to become white hot.

Previously we concluded that the combustion of carbon called forth phenomena which reminded us of a sunrise. In contrast, the glaring glow of the magnesium flame is comparable to the direct, full glare of the sun, high above the horizon, in a very clear sky.

It is very interesting to study which "technique" nature uses to produce this glaring white magnesium light. The heat of combustion of magnesium is very large, so a great amount of heat is evolved. For the most part, this heat is passed on to the particles of magnesium oxide.

These are very white and consequently do not radiate a lot of heat. Furthermore, they are very easily heated because of their low specific heat. As a result the temperature can quickly rise to a very high level.

In every respect there is something violent in the essence of magnesium, which expresses itself in large contrasts and sudden, violent manifestations. The lack of smooth transitions also expresses itself in that we do not find any color in the flame or in the compounds of magnesium.

In this context the properties of the products of combustion are also very odd. Magnesium oxide is only slightly soluble in water and very difficult to melt (melting point 2,800°C). Thus, on its own or mixed with other substances it is the best fire-proof substance available.[50] The combustion of magnesium is not only associated with this dazzling lighting-up, but also the oxide is thrust deeply into the realm of the earth element. More than other elements, magnesium is caught between the polarity of sun and earth. The fierce flaring-up in combination with the formation of a substance which is very insoluble and hardly fusible, is an image of the contrast between sun and earth.

Something similarly forceful expresses itself in the extraordinary power of reduction, but we will speak of that in detail later.

Phosphorus as Representative of the Stars on earth; Results of a Comparison with Magnesium

Previously, by using a striking example, we have demonstrated how an element, which has been severed from a part of nature can, under certain conditions, call forth images of that part of nature. If we apply to phosphorus this rule, then the fact that the experiments always give rise to an image of the starry heavens demonstrates: phosphorus must be a substance which has a special relation to the stars. We may also assume that phosphorus receives influences of the stars and transmits them to living beings in the same way as magnesium does with the sun's influence. From the viewpoint of contemporary science, this may be regarded as very unlikely. However, in the case of photosynthesis the case is clear. There we are dealing with chemical reactions and processes which are energetically easily understood. In the case of transmission of starry influences, we are dealing initially with form-bringing influences. Such processes are currently very difficult to substantiate. Therefore, it seems important to find experimental methods for this.

Also, from a different approach, we may assume that something may be found experimentally. In general, life inaugurates processes which are the opposite of those occurring in inorganic chemistry. In the case of magnesium, we see that the same substance which radiates light in such a magnificent way, is used to catch and process the mighty sun rays in chlorophyll. We may also expect that the substance which is repeatedly subject to such delicate influences of the stars, may itself be helpful in catching the influences from the stars on living beings.

Whereas the power of the sun, caught via a magnesium compound, activates right down into the metabolism and material formation and the spatial extension of a plant, the influences of the stars caught by phosphorus initially causes contraction and rigidification and the appearance of delicate structures, orientated towards the center. That follows at least from the particular appearance of phosphorus at places like our skeleton, nervous system, in seeds of plants, and nuclei.

Once we have understood something of the relation that exists between phosphorus and the stars, then we can easily understand that the glowing of even a piece of phosphorus expresses this. If we leave a piece of phosphorus lying about, it will very slowly disappear while emitting light. We could say it disappears as "cold light." That is a property which is foreign to earth-like substances, but which does belong to the stars. To our perception, a star is a point-like light source, thus, something whose sole property is the capacity to radiate light.

Looking back on what has been discussed, the fact that just the substance which dives into the earth element, yet clearly points to its relationship with the stars, is also the substance which guides man's spirit into the material world, appears very important. That is a connection we should ponder a lot.

Magnesium Chemistry

In the chemistry of magnesium, the relationship to oxygen comes strongly to the fore.

The power of the magnesium flame is predominantly a result of the affinity of magnesium for oxygen. In a certain sense, the fact that magnesium is not attacked by oxygen at normal temperatures seems to contradict this. However, magnesium protects itself with a thin layer of oxide. The oxidation at higher temperature proceeds more quickly and more violently than in the case of other metals; the lower-temperature oxidation, which is typical for metals, becomes suppressed.[51] We see from that that magnesium is characterized by marked contrasts.

By removing the film of oxide, for example with iodine, magnesium reacts strongly with water at normal temperatures:

$$Mg + H_2O \rightarrow MgO + H_2(g)$$

Magnesium has such a strong reducing power that it can be used to reduce nearly all metal oxides. The following experiment is particularly nice. In an Erlenmeyer flask, we bring water to boiling. Then, we ignite a piece of magnesium ribbon and bring the burning end through the neck of the flask in the atmosphere of water vapor. The flame becomes even more intense, if possible. Moreover, a barely visible hydrogen flame appears at the opening of the flask.

Also, it is very amazing that it is possible to ignite a mixture of magnesium and heated silica earth, in the ratio 2 to 1, with just a lighted match.

The mixture burns as if silicon oxide were an oxidant.

$$SiO_2(s) + Mg(s) \rightarrow 2\,MgO(s) + Si(s)$$

Even sand, which is more dense, can be reduced by magnesium.

A mixture of magnesium powder and solid carbon dioxide may be ignited with a piece of burning magnesium. It will burn brightly, giving off carbon soot.

At first sight it seems as if magnesium oxide exhibits very little chemical activity. It can behave as a very inert, inaccessible substance. It closes itself to all possible influences. An example which we have already come across is that being the whitest substance, it reflects all light. Heat also does not have any point of attack. With normal means it can hardly be brought to melt (melting point about 2800°C), and it resists boiling most strongly (boiling point 3600°C). If we heat it strongly, its density and imperviousness even increase. In this way we can produce a substance which is fire-proof and resistant to chemicals. It is used to make crucibles. Further, it is used as holder for incandescent mantels. It may even be used instead of platinum wire to hold salts in the flame when investigating them on the color of their flame.

Mixtures of magnesium oxide and certain other substances are used to make fire-proof stones for electric ovens and similar.

If we let some magnesium oxide which has been freshly prepared by burning magnesium ribbon, drop into boiling water, it will hardly dissolve, but the little that does dissolve is strong enough to turn litmus paper blue.

The strong acids easily dissolve magnesium oxide. The magnesium salts of the strong acids are neutral, from which we may conclude that magnesium hydroxide is a strong alkali.[52]

Due to its special properties magnesium oxide is especially suited to remove excess acid from the stomach. We can even use it to neutralize strong acids which have accidentally gone into the stomach. It will quickly dissolve until the acids have been neutralized and then remain there quietly, without attacking the mucous membrane of the stomach as other alkaline substances do.

In the phenomena discussed so far, we always meet something deeply embedded in the character of magnesium. Under normal circumstances both the metal and the oxide are substances which exhibit a certain inertia and inactivity. But as soon as they enter a realm where strong forces reign, then they also develop a strong activity and great force.

Phosphorus in Mythology and Industry

We regularly find mythological names in chemistry. That is a peculiar phenomenon in a realm which stands in the bright light of modern scientific thinking. Even more odd is that these names often correspond to the nature of the substance. Phosphorus means "light bringer" (Greek foos foros). It is also the name of a spiritual power, which is normally called Lucifer, which means "light bearer." Prometheus is a Greek name for the same being.[53] According to tradition, he stole fire from the heavens and brought it to the people. In this connection various facts are very remarkable. In the first place images of the "heavenly fires" are produced continuously through phosphorus. Further, in times of peace phosphorus is used widely in the production of matches, that is – to light a fire. The Dutch word for matches is "lucifers." The ignition of these "lucifers" is based on the presence of phosphorus.[54] The mythological names can penetrate even into the sphere of industry. At the same time, we have here an example that if industry is really well developed, the essence of a substance becomes apparent.

War always brings about a culmination in industrial activity. Especially during World War II, during which a specially advanced chemical technology developed, the nature of phosphorus was made apparent in quite a remarkable way. It was used to produce "fire from the sky" – a mixture of phosphorus and sulfur. The substance is liquid and ignites at normal temperature. Extinguishing it does not help. We see that also here, where we are dealing with an exceptional piece of technology, phosphorus is being used to create fire, thus for the same purpose as in the manufacture of matches. With the aid of the "Fire from Heaven," fire is ignited on earth.

According to tradition, the heavenly fire stolen by Prometheus was not only used by human beings, but also indicated an awakening, a first enkindling of the human spirit. Prometheus, thereby, gave to humanity the possibility of being creative both in a technical and spiritual sense.

The fact that we carry the capacity to take initiatives in us, amongst others, places us again and again at a beginning. That we have images of the power of initiatives is due to the fact that phosphorus or some specific compounds can spontaneously ignite. Our power of initiative acts again and again as the fire which was originally ignited.

From all that has been said, it can again become very clear how phosphorus is particularly able to be the physical basis for our power of initiative.

Nitrogen

Nitrogen as a Component of the Atmosphere— An environment for Living Beings

In the chapter on oxygen we have already partially characterized those gasses which are part of the atmosphere, as far as mobility, density, and transparency to various radiations are concerned, and remarked that they were "all present." That, of course, applies primarily to nitrogen, since about four-fifths of the atmosphere consists of nitrogen.

Of all the properties of the atmosphere that surround us, we can divide these into a number of positive and also a greater number of negative ones. We shall see that especially in the case of nitrogen, the negative properties are at least as important as the positive ones. For positive properties we can count: becoming visible when in strong movement, exerting pressure, promoting or supporting and maintaining many processes, especially respiration and combustion. For negative ones, we may indicate its invisibility due to its transparency and colorlessness, the lack of form, smell and taste and low density; because of these "negative" properties, the atmosphere is hardly visible. If we remove oxygen from the atmosphere, then some of the important remaining positive properties disappear: the basis for respiration, the stimulus for and maintaining of combustion. Nitrogen (as an atmospheric gas) is hardly accessible to chemical influences.

We find a noteworthy contradiction in the occurrences of nitrogen: it is a substance which comes into prominence in our environment, yet it also conceals itself from our senses. That gives rise to something mysterious in the way it appears. At any rate it may be said that to a very high degree nitrogen is typical of the atmosphere. In many ways all other components of the atmosphere show similarities to nitrogen. In the course of our discussions we will also come across the converse, where more than other substances in the atmosphere, nitrogen shows a connection to the properties of the atmosphere, in the whole way it expresses itself.

Perhaps this is the appropriate point to mention that we have here one of the rare occasions, or even the only occasion, where a substance which mainly occurs in elemental form plays such a vital role.

We normally take too little notice of the fact that the gasses found in the atmosphere are the mediators for an extraordinarily varied interplay of light and color. The atmosphere is a realm where colors continuously emerge and disappear to a much greater extent than anywhere else in the world. If the atmosphere around us did not exist, then the sun would shine from a black sky like a scorching, glaring light source. Reducing the glare to a certain mildness, mixing the bright with the dark to produce the blue of the sky, creating the yellow brilliance of mid-day, and the red glow of a sunset, creating the wonderful sparkling of colors: all this is brought about by the atmosphere. If in this way a truly magic palace has been built for light, then nitrogen is intimately involved in that process.

We can make the impression of these phenomena of nature more explicit through the following comparison. On a brilliant sunny day we let a large quartz crystal sparkle in the sun. We then immerse ourselves into its inner space, filled with cool beauty. These phenomena stand in sharp contours before us, and we experience ourselves completely separate from what we are seeing. In that manner we acquire a clear, conscious relationship to a very noble aspect of the element earth. Afterwards, we observe the atmosphere. Then we can feel ourselves one with an infinite space full of a marvelous glittering and radiance. This experience can give a deep impression of our relationship with the element of air. This completely surrounds and engulfs us. It is the real, proper element in which live.

Certainly, air is the environment for light, but least of all is it the carrier of light. This becomes apparent when we isolate part of the air from the atmosphere. We are then left with something which is as devoid of light as we can imagine. A gas, isolated for scientific or technical reasons, really has entered a schematic or shadow like state in comparison to its condition in the atmosphere. In the case of nitrogen, this absence of light has gone so far that during combustion, which takes place at very high temperatures, it absorbs heat instead of evolving heat and light.

We take it for granted that nitrogen is colorless and transparent—as are the other gases of the atmosphere. However, we should be very surprised! Frequently, the opposite situation occurs. Storms often create dense clouds of dust; volcanoes eject opaque, pungent vapors; the water vapor of the clear sky can condense to a slight opaqueness or turbidity, or else a dense, opaque mist. Fortunately, all this disappears again, and we then experience what we were missing. Only after all the conditions that made things opaque are gone can the influence of the sun fully reach the earth. Furthermore, only now can we really rejoice in the way our gaze can roam freely through the atmosphere and in the wonderfully broad view in it. The more the turbidity and vapor disappears, the clearer the image of the things around us and of the heavenly bodies becomes.

Thus we become aware how essential the transparency of the atmospheric gasses is for the relationship of the earth with the extra-terrestrial, and primarily with the sun; likewise this is true for our consciousness, because it makes a precise observation of our surroundings possible. If the atmosphere had been denser, then the clarity of our mental image life, which depends to a large extent on the sharpness of our optical images, would not have reached such heights.

However, not only in this way, but in a much more direct way, the atmosphere contributes to the fact that we are surrounded by the wonderful, amazing play of forms, which we encounter in the animals and plants around us. Inherent in the atmosphere is the urge to expand. In vicinity of the surface of the earth there is hardly ever enough space for such expansion, because there is too little free space. The atmosphere then oscillates around the point of equilibrium between the urge to expand and pushing-back. Thus, atmospheric pressure comes about. The urge to expand brings the atmosphere there where there is nothing else. As soon as an emptiness is created somewhere, air plunges in; if a thing needs more space, then air moves aside. We must imagine that a growing plant not only develops positive forms, filled with plant matter, but also negative ones in the atmosphere. As in the case of plants, so it is with all created things, be it a stone, animal, or human being; the atmosphere always creates a negative air-form about it.

This behavior of the atmosphere is of great importance for the behavior of water. As soon as water is permeated by warmth, we see a tremendous urge to expand, which only manifests itself when the air above the water has been taken away (the pressure above is reduced); then, the water starts to effervesce strongly and evaporates. If we only eliminate the moisture of the air, while the pressure remains constant, then a stronger evaporation occurs, although this only takes place at the surface and proceeds quietly. Thus, we observe that the urge of water to evaporate is actually reduced by the presence of air.

This same phenomenon is also of paramount importance to living organisms. If we bring them into a space where the pressure has been greatly reduced, then they are severely harmed because of the frothing of their body liquids. Every plant may be likened to a delicate water column, surrounded by more solid parts. Human beings and animals even carry an amount of liquid with them, which is not totally enclosed by their skin. If these beings were not surrounded by the atmosphere and their body liquids kept in check by air pressure, then the multitude of forms we experience would not be possible.

Again we must be aware that in these processes, nitrogen plays a major role.[55]

We can further reduce the amount of nitrogen in the air, without reducing the pressure. We will then notice that all fires will burn more vigorously. As the nitrogen content is increased, combustion is reduced. From this we can become aware how nitrogen in the atmosphere regulates the influence of oxygen by thwarting its action.

In this way we arrive step by step at an image of the importance of nitrogen in nature.

That nature presents itself with such a richness is mainly the result of the properties of nitrogen. Not only does it enable us to see everything around us freely, but it also contributes to the decorative colors of the sky, and it offers living beings the opportunity to appear in those characteristic, finely differentiated forms. Nitrogen gas acts as if it were an efficient, prudent director. In a certain sense, it pushes itself forcefully into the foreground, accompanies and surrounds everything that happens. It, thus, tempers a series of processes which otherwise would proceed too explosively; it, thus, contributes to maintaining an equilibrium. On the other hand, it is always prepared to get out of the way and retreat. Every being which wants to move and express itself is given every opportunity; every process may develop. While making itself invisible, it lets everything else appear as magnificent and expressive as possible.

Nitrogen in the Soil

In general there is little nitrogen in the crust of the earth. We could say that nitrogen is badly adapted to the earth element. It exists only in the soil layer which is in close interaction with vegetation, where we find a substantial percentage of nitrogen.

While atmospheric nitrogen in various ways seems to exert a moderating influence, this soil nitrogen seems to act exactly opposite. In Holland, we are only too familiar with these nitrogen-poor soils, where mostly heather grows or where peat moor is formed. There we find a one sided, meager vegetation. Often we encounter insect-eating plants, e.g. sundew, Fettkraut, and bladderwort. Actually, these are plants which do not fully connect to the soil, but seek a relation with the atmosphere in an unusual way. In a different manner that is also the case with the genus Genistra, which is capable of processing the nitrogen from the air directly with the aid of bacteria which are situated in root nodules. Even on the poorest soil we can find these plants come to an exuberant flowering. However, life on a nitrogen-poor soil will generally not thrive. If we want to enliven such a soil, then the first step is to enhance the activity of nitrogen, which can be achieved by planting legumes such as lupine or vetch.

Only since the end of last century do we know that a fertile soil has its own nitrogen cycle. If a soil is deficient in nitrogen, then it is absorbed from the atmosphere and converted to all sorts of compounds. However, if there is an excess of nitrogen compounds, e.g. in places where excrements have accumulated or where a corpse is decomposing, then nitrogen disappears as a gas or as ammonia gas.

The fact that nitrogen vigorously stimulates life in the soil is really something very surprising. The substance which in the atmosphere dampens the activity of the great life bringer—oxygen—forms the basis for the substance which makes life in the soil possible). This dampening originates in nitrogen, when it is in elemental form and not associated with oxygen. The enhancing of life originates especially in compounds of nitrogen and oxygen. It seems as if nitrogen directly enhances oxygen-activity.

In the fact that nitrogen behaves in such opposite ways in soil and atmosphere, we meet for the first time an example of something that will always catch our attention: Nitrogen can manifest itself in various areas in two contrasting tendencies. It always moves between two extremes.

Nitrogen in Protein and in the Atmosphere

When looking for substances in living beings from which it is possible to analyze nitrogen, then the proteins present themselves as the most important. More than other substances, proteins can be seen as the true carriers of life.[56] The livelier the protein, the less formed and the more changeable it is. We can observe this very clearly in the protoplasm of plant cells, which consist for the most part of proteins. We can often see it move or stream, but in any case it participates in a vigorous chemical transformation of substances and, more importantly, builds the solid, structural parts of the cell. Upon analysis, it is possible to obtain elemental carbon, oxygen, and hydrogen from nearly all organic substances. In the decay products of proteins, the liveliest, most restless, mobile substance we find is nitrogen. This indicates a relation between mobility and the presence of nitrogen. This also agrees with the fact that in animals, where mobility is typical, we find a higher percentage nitrogen than in plants.

Now it is very surprising to find that analysis of proteins yields the same elemental substances as in the atmosphere[57] - carbon, hydrogen, oxygen, and nitrogen—and also a small amount of sulfur. In those places in the living organism where nitrogen plays an important role, we are reminded of the atmosphere. Thus, in his body, like other living beings, the human being carries the same substances that surround him, even in crucial parts of his body.

But in which of the various conditions do the substances occur? Inwardly, they are sustained by the driving forces of life that build whole forms; out in the atmosphere, they are so dissociated from the form-creating forces, that even as chemical elements they can mostly exist independently and separate from each other. In its composition, the atmosphere reminds us of proteins, but proteins that have totally lost the typical protein quality, e.g. being the starting-point for constructive forces. After all that has been discussed, it should not be surprising that sulfur is not found as a constituent of the atmosphere.

It is characteristic of the substances which form proteins, that in nature they are found as elements. They, thus, distinguish themselves from the other elements which belong to the constituents of living beings and which are almost exclusively found as compounds, hardly ever in elemental form.

In conjunction with the other substances which also occur as atmospheric gases, we have now found nitrogen in the depths of our metabolism in a state of intensive association with the formative vital forces. That corresponds completely with the occurrence of nitrogen in the soil.

What is the significance for the human organism of finding nitrogen in the atmosphere? Initially, it could appear as if it were neutral or even hostile towards the processes occurring in the human body, because it arrests the life-bringing action of oxygen. However, it we reflect more thoroughly on the matter, then we will realize that it is just as important for the sense organs as the bound nitrogen is important for the metabolism. The same remark may be made for the other atmospheric gasses, in so far as they resemble nitrogen. As far as the eye is concerned, we have already formed the basis for this statement when we spoke about nitrogen as being the "director" or "stage manager." Our eye is inconceivable in its present form and function in the absence of the transparency of the air. The gasses of the atmosphere and especially nitrogen are also the medium for sound. And, heat radiation is transmitted nearly unhindered, so that we can feel the heat of a distant fire, for example. And lastly, because air does not excite our organ of smell, it can carry aromas to us.

In every living being there occurs a kind of meeting of living and dead nitrogen. Living nitrogen in protein takes part in the formation of ever new forms. Dead nitrogen plays around these forms and envelops them, holds them together. Moreover, the nitrogen which surrounds human beings and land animals serves as a field for their consciousness, in as much as this rests on the activity of the senses. We have again met nitrogen in two opposing states: nitrogen that takes part in the expression of our consciousness, and nitrogen which assists in building our organism. In the first case, nitrogen is in a state where the form-creating forces of life are missing; in the latter, where these form-creating forces are active, consciousness is absent.

For our ability to use our consciousness, we can thank the fact that there is a sphere where substances have both somewhat let go of the form-creating forces and have fallen somewhat.

Nitrogen in Nature between Oxygen and Hydrogen

In fertile, well-kept soils nitrogen compounds have the tendency to covert to the highest oxidized state, to the nitrate state (viz. KNO_3). Both ammonium compounds and nitrites (e.g. KNO_2) are oxidized under favorable soil conditions. Under these conditions soil is able to take up and process nitrogen from the air. These processes take part with the aid of bacteria.[58]

In certain exceptional cases, these processes can be so vigorous that saltpeter (potassium nitrate) emerges from the soil. That can occur in areas where the soil contains a large amount of chalk or potash. In South India, this phenomenon occurs during the transition from the monsoon season to the dry season. If we touch such a salt layer with a glowing piece of wood, then we will see this flaring up. That indicates a vigorous activity of oxygen.

Under certain less favorable conditions, reduction may occur in the soil, whereby nitrates are reduced to nitrites and even may be converted to ammonia. The formula for ammonia NH_3 indicates this reduced state particularly clearly. Not only does this compound contain no oxygen, but rather it contains hydrogen in considerable amounts. Hydrogen is one of the easiest substances to oxidize. If we come across an accumulation of nitrogen-rich substance, such as horse dung, we can often become aware of the smell of ammonia. This is also quite noticeable on the Guano Islands off the coast of Chile.

When a soil produces nitrates, then that means an increase of the fertility of the soil or, expressed differently, an enlivening of the soil. Nitrogen is brought more into the sphere of the life-bringer, of oxygen, and is, therefore, being activated. Conversely, oxygen and life in the soil are stimulated by nitrogen. It really is obvious that plants can take up nitrogen best as nitrate, the state where life is brought nearest by oxygen.

For example, as soon as gaseous ammonia is formed by manure or an animal corpse, an impoverishment of nitrogen compounds occurs. Nitrogen combined with hydrogen is removed from oxygen, and, thus, from life.

Nitrogen Chemistry

The chemistry of nitrogen is very varied. However, when we have an overview it is possible to distinguish two main types of compounds: those derived from ammonia, and those that originate from nitric acid or nitrogen-oxygen compounds. Here also we find nitrogen oscillating between oxygen and hydrogen.

We will first of all turn our attention to those compounds which have oxygen and nitrogen as their basis. The most important are nitric acid and the nitrates. Nitric acid (HNO_3) should really be called nitrogen acid, if the way chemical names were assigned was used consequently. Its basis is nitrogen which has been activated to the highest degree by oxygen and water. The anhydride is nitrogen pentoxide (N_2O_5), the oxide with the highest oxidation number. Nitric acid is an extraordinarily active acid. There are three reasons for this:

       1. Its salts are all soluble.
       2. It is a very strong acid.
       3. It acts as a strong oxidant.

That its salts are all soluble is really self evident. Even as a salt nitrogen, which is in the first place an atmospheric gas, does not completely rigidify. We encounter also here that nitrogen hardly binds itself to earth.

Being a strong acid, nitric acid gives a rather dramatic quality to nitrogen chemistry. Chemically, nitrogen gas is exceptionally inert, and so the great activity of nitric acid is, therefore, even more distinctive. The oxidative power of nitric acid is so strong that it can dissolve noble metals like copper, mercury, and silver.

Regarding its composition, we could call nitric acid "air acid," because through analysis we can isolate the three most important atmospheric gasses which occur as chemical elements—hydrogen, oxygen, and nitrogen. Because it is combined with oxygen, nitrogen as nitric acid can participate in every conceivable reaction and form all sorts of compounds. At the same time, it brings with it something mobile and changeable, which belongs to the essence of air. On the other hand, oxygen becomes activated by nitrogen, so that at normal temperatures it is much more strongly oxidizing than is atmospheric oxygen. Moreover, nitrogen acts as a forerunner of oxygen. With the aid of nitric acid, it is possible to introduce oxygen in such a way in the inner structure of a substance that we can speak of an internal oxygen supply. As a result, some of these substances acquire explosive properties: an internal combustion could start. We can easily demonstrate this with gun-cotton (nitro cellulose, pyroxylin). If we impregnate a tuft of cotton with fuming nitric and concentrated sulfuric acids according to a 2:1 recipe—a process called nitration—we get something which only seems to differ from cotton wool in being somewhat rougher. If we ignite normal cotton wool, we get a small flame while the mass slowly carbonizes and glows for a long time. But if we ignite a piece of nitrated cotton, then everything is over in an instant. The whole mass disappears in a large flame, which flashes out to all sides. Nearly nothing remains but a mixture of gasses consisting of nitrogen gas, carbon dioxide, and water vapor. The cotton wool has literally "vanished into thin air." A process of combustion which really belongs in the realm of the air, has occurred in the realm of earth. We could also say that this explosion is a protest against the earth-element as such.

In these processes we can discover a parallelism to the relationship of nitrogen to the living soil. Nitrogen there is also a promoter of oxygen activity, but in such a way that life is enhanced.

The oxidative properties of nitric acid have been totally given to the salts of nitric acid: the nitrates. The salt most commonly used in this connection is potassium nitrate or saltpeter. With it we can do especially beautiful experiments. For example, we can fix a medium test tube filled with potassium nitrate to a laboratory stand. Then, we carefully heat the tube with the Bunsen flame. When the salt has melted, we drop in a piece of glowing charcoal. As soon as it touches the surface, it flares up vigorously and jumps about. If we similarly drop in a piece of sulfur, then it flares up brilliantly and the tube shines like a lamp as a result. After a while, the tube becomes so hot that the glass melts and the bottom part loosens itself and drops down. If we do the same experiment in an iron spoon, than we can let a piece of wood be devoured by the glowing mass. Viewed from an alchemical viewpoint, we have the odd phenomena that a salt resembles the fire element; thus we encounter sulfur-activity in the sphere of salt.[59]

From time immemorial, a mixture of saltpeter, sulfur and charcoal has been used as gun powder. Furthermore, saltpeter is used to ignite fireworks. Saltpeter has, thus, always played an important role in the pyrotechnics and the explosives industry.

We have already discussed how all nitrates are very soluble in water. In this case we may not conclude that there is a special kinship with water; however, the following does apply: inasmuch as the nitrates are crystalline, they are part of the earth element. At the same time, as nitrogen compounds, they are a protest against the earth element. In this case, the solubility in water means above all a freeing from the earth element. In the case of potassium nitrate, the usual temperature-dependent solubility increases much more rapidly. This indicates a double kinship with the warmth element, which also emerges in the experiments with fire. In the case of potassium we showed previously that there was a certain kinship between it and fire, and we also see that for the nitrates.

As far as the curriculum is concerned, I still think it is justified to start with the nitrates and especially with "chili saltpeter" (sodium nitrate $NaNO_3$) and saltpeter (KNO_3). The students already know these substances, and, moreover, it is the historical sequence in which humanity got to know and produced nitrogen compounds. And just as importantly, it is the natural way.

Indeed, nowadays we use atmospheric nitrogen to produce nitrogen compounds so that in the end we produce nitrates by way of nitrogen oxides and nitric acid, but notwithstanding the fact that this is also the avenue followed in nature, this pathway plays a smaller role in the world.[60] It is possible to convert saltpeter to nitric acid and convert this to nitrogen oxides. If we then discuss the technical processing of atmospheric nitrogen, then the structure of these substances in particular becomes clear.

If we take the other pathway and form the various compounds starting with nitrogen, we will achieve a very clear picture, but we will be in danger of remaining sketchy, whereas the characteristic, peculiar properties of the various compounds do not show up.

In general, we may say that nitrogen is always brought down to earth by oxygen. Its oxides, if they are gaseous, have a greater vapor density than atmospheric nitrogen. Some are liquids, and one occurs as a solid.[61] Nitric acid is a liquid. All nitrates are well-crystallized salts.

Under the influence of hydrogen, nitrogen has the tendency to flee the earth. That is expressed in the first place by the vapor density of ammonia (NH_3), which is a little less than half that of air.

Ammonia is one of the substances formed when tissue decomposes. In the past, ammonia was obtained from decaying human and animal urine. Subsequently, ammonia gas was obtained in large quantities as a result of cleaning crude coal-gas. And here we are again confronted with a product which is the result of decay and fossilization of plant matter.

Ammonia is exceptionally soluble in water. That by itself is already surprising, but even more surprising is the fact that this ammonia solution has fairly strong alkaline properties. It is assumed that ammonia and water are converted to an unstable alkaline ammonium hydroxide (NH_4OH). Just as easily as it is absorbed, it can volatilize out of the water. By adding a little strong acid to the solution, the strong smell disappears. Ammonia is then retained as an ammonium salt, e.g. ammonium sulfate [$(NH_4)_2SO_4$].

These facts become even more remarkable when we consider that ammonia is synthesized from the element which makes everything volatile, transparent, soluble, mobile. These are not the most suitable properties to make a solid base. The typical alkaline substances (bases) are opaque, insoluble, and heavy. The few bases which are very soluble and some very few which are even easily melted, are certainly not very volatile. Ammonia gas, because it is gaseous, has more similarity to acids than to alkalies. If we mix crystalline ammonium sulfate with solid sodium hydroxide and then moisten the mixture a little bit, then gaseous ammonia will be formed. This process is based on an alkaline displacement, which proceeds like an acid displacement.

The character of nitrogen emerges particularly clearly in these curious contradictions. We encounter here again how nitrogen expresses itself in two opposite conditions. Nitrogen would not be nitrogen if another nitrogen compound with at least fairly pronounced alkaline properties stood opposite the strong nitric acid.

If we blow ammonia gas into a gas flame, it will burn inside it with a small yellow flame. We are again confronted with opposites in the chemistry of nitrogen: nitric acid and its associated substances have oxidizing properties. Ammonia is combustible and brings reducing properties to its compounds. However, the oxidizing properties of nitric acid are very well developed, whereas ammonia only burns with difficulty.

Something similar emerges when we compare the strength of the alkaline properties of an ammonia solution with the strength of the acidic properties of nitric acid; although nitric acid and ammonia are opposites, they do not hold each other in balance. Nitric acid always has the upper hand, as it were.

It is possible to create very powerful basic substances with the aid of ammonia, the so called amines. We must substitute part or all of the hydrogen of the ammonium-group [62] with hydrocarbons. It is interesting to note that with increasing base-strength, the combustibility increases. Until now, we saw that bases came about through rusting, and acids through combustion; here we have easily combustible substances.

That the ammonium salts have very remarkable properties will have become clear from what has been said. When heated, sal-ammoniac (ammonium chloride) volatilizes as a white smoke. Ammonium carbonate volatilizes already at room temperature and decomposes into ammonia gas, water vapor, and carbon dioxide. Ammonium sulfate produces gaseous products when heated. These phenomena are a consequence of the fact that the "basis" of these salts cannot give them a solid foundation.

All ammonium salts are easily soluble.

From out of all of this, the striking kinship of the ammonium salts with air may be deduced.

Ammonia gas is combustible to a certain extent. If it forms a salt then it imparts to the compound the ability to be combustible. Thus, very interesting phenomena should be shown by ammonium salts derived from acids with oxidizing properties, such as chromic or nitric acids. Especially remarkable is the behavior of ammonium dichromate. Make a little mound of the pulverized orange-red crystals on a fire resistant plate and ignite it with a gas flame. With a sizzling it starts to burn, showering sparks around; at the same time it grows to a loose gray-green mass of chromium(VI) oxide. It strongly reminds one of an erupting volcano.

Ammonium nitrate (NH_4NO_3) and ammonium nitrite (NH_4NO_2) are both very remarkable substances. Normally a salt is the result of the reaction of an acid and a base, which relate to a combustible non-metal and a rusting metal. A salt is the harmonizing of two opposite states of substances. Ammonium nitrate is also the result of an acid and a base, but these are both derived from the same substance. Nitrogen forms both acidic and basic substances, and seeing how nature is organized, together they must produce salts. However, these are not in a well-balanced, peaceful state like other salts. No, so often in nitrogen chemistry things are opposite to what they are normally. Ammonium nitrite is very unstable and volatilizes at the slightest cause. Ammonium nitrate burns when heated and can be used as an explosive. To a degree, these salts stand under strong tension, whereas the chemical element from which they are derived is very well-balanced and not reactive.

We can show these properties with the aid of some very impressive demonstration experiments. For example, we can melt some potassium nitrite on an iron tile and throw some ammonium chloride on it from time to time. A flaming effervescence occurs, and we experience a miniature thunder storm. Ammonium nitrite, which resulted through a double reaction, decomposes to water and nitrogen. If we drop some ammonium nitrate on a hot iron plate, then it will decompose accompanied by flames.

In an actual explosion, which results from still higher temperatures or detonation by mercury fulminate, the salt is converted to nitrogen and oxygen gasses and water vapor, thus into the constituents of air.

$$2\,NH_4NO_3(s) \rightarrow 2\,N_2(g) + 4\,H_2O(g) + O_2(g)$$

In conclusion, it is a very impressive experiment. The action of nitric acid can often be very violent. We can, for example, pour some dilute nitric acid over some pieces of zinc in a small bowl. Immediately the mass effervesces and dark brown vapors (nitrogen dioxide NO_2) are produced. If we place a large cylinder glass over the bowl, it quickly fills itself with a dark brown transparent gas. If, as soon as the reaction has come to an end, we then pour some ammonia solution on the table around the cylinder glass, then white, opaque clouds are formed in the cylinder glass. If we let these carefully escape through a crack, then a heavy, milky mist will move across the table, forming the most wonderful vortices.

Aluminum

Aluminum as Element

For the discussion of aluminum, although it does not appear anywhere as an element, we may just as well start with the metal itself, because it is very well known and has very characteristic properties. In the first place, in comparison to the "heavy metals" which were the only metals known previously, its low density is very remarkable. Its specific density is comparable to the average specific density of rocks, for example, marble. Because aluminum, therefore, belongs to the group of light metals, it should have a chemical activity comparable to sodium, potassium, and the like; that is not very apparent, at first sight. It is conspicuously inert under the influences from the atmosphere. Aluminum is used as a powder in paints for lantern posts, fences, and balloons which are then covered with a protective, matte shining metal layer.

Aluminum may be manufactured into a light but resistant material, which is ideally suited to be used in making airplanes and in high-speed vehicles. In the aircraft industry, its lightness is especially important, and for high speed vehicles, its small mass allows for quick braking and quick acceleration.

Aluminum is also used quite extensively to manufacture containers, small household utensils, and scientific instruments.

The attempt has been made to use aluminum for ship hulls; however, a surprising difficulty was encountered: aluminum is rapidly attacked by salt solutions.

The same property must be taken into account in the manufacture of cooking pots. Many previously manufactured pots have become unusable after salted food was left standing in them. However, by alloying aluminum with different metals, this problem has largely been overcome.

Chemical Properties

To demonstrate the above property, we can do the following experiment: Set out three beaker glasses, pour in A dilute hydrochloric acid, in B a solution of sodium chloride and in C some sodium hydroxide solution. Then add some aluminum powder, from which fat has previously been removed by boiling it in alcohol. The liquid in A begins to effervesce immediately because hydrogen gas is evolved; aluminum powder dissolves. This should not surprise us, because it occurs with other metals, also. What takes place in glass C, however, is very astonishing. Taken at face value the same occurs as in glass A. Aluminum powder dissolves just as well in sodium hydroxide as in hydrochloric acid.

It is now also understandable how a substance which can dissolve in hydrochloric acid and in sodium hydroxide will be attacked by a solution of sodium chloride. This process is only much slower than in the case of the acid and the base.

In these experiments we encounter a distinct characteristic of aluminum. It is a metal which does not place itself only on the side of the bases and, thus, is not only attacked by acids, but it is accessible to the influence of both acids and bases.

Strangely enough, aluminum is not attacked by some acids. For example, nitric acid and sulfuric acid hardly attack it at normal temperatures.

For this passivity of aluminum, a very thin, transparent and, therefore, invisible oxide film is responsible. This oxide film forms as soon as unprotected metal is exposed to the atmosphere. We can remove this film by rubbing the metal with mercury or mercuric chloride ($HgCl_2$). It is even easier using a drop of water and a few granules of mercuric chloride. This causes some mercury to form. By careful heating we can rid ourselves of the water. Now a very strange process starts. It seems as if a thick, fiber-like mold is growing on the metal. That can happen so quickly that we can follow it with our eyes. We have here a very fast process of rusting. Aluminum oxide (Al_2O_3) is formed.

If we had dipped the same piece of "activated aluminum" in water, it would have reacted strongly and hydrogen gas bubbles would have arisen.

Aluminum powder, when scattered into a flame, burns to oxide while producing fairly glaring bands of sparks. Aluminum foil burns with a glaring light.

When mixed in the correct proportions with oxides of heavy metals, aluminum powder can act as a strong reducing agent. By igniting such a mixture, the heavy metal is set free, while an extremely violent incandescence is produced. In the past a mixture of iron oxide and aluminum was used to weld tram rails. It has also been used to make fire bombs.

From such experiments, we may conclude that aluminum as an element is charged with tension and energy.

Aluminum oxide, as it is produced from aluminum metal with the aid of mercury, could be referred to as a substance without character – it is without form and color. It dissolves well in acids, and nearly as well in bases. With acids, it forms aluminum salts; with bases, aluminates, e.g. salts of alumina acid. This behavior of the oxide is called "hybrid" (zwitter-like) or amphoteric[63], by which we mean that it behaves as a base with respect to acids, and as an acid with respect to bases. Its behavior is thus determined by the circumstances, not by the substance itself.

As soon as aluminum oxide is heated strongly, it loses its susceptibility to influences and its malleability. If we hold the cone of a fire resistant small rod in the intense glow of an oxyhydrogen flame and then blow in some aluminum oxide powder, then it can even condense to gemstones, to corundum, ruby, sapphire. The hardness is caused by the aluminum oxide, the color by small admixtures of metal oxides. From this process we can see that fire drives aluminum oxide deep down into the earth element. It is very surprising that a substance which at first seemed so perfectly passive should suddenly retain very forcibly everything of the form and color to which it has been subjected.

We will always come across such phenomena when dealing with aluminum compounds.

The malleability, alternating between acid and base, becomes very evident when we start with an aluminum sulfate solution, and first add some sodium hydroxide and then some hydrochloric acid. If we do not add too much sodium hydroxide solution, then initially a jelly-like precipitate (gel) forms, which re-dissolves with an excess of sodium hydroxide solution. It can be recovered again by adding an acid-like hydrochloric acid. Additional acid dissolves it again. The resulting precipitate always has the same composition: aluminum hydroxide or $Al(OH)_3$. Not only can it react with acids but also with bases, to form salt-like substances. This alternating behavior may be expressed as follows:

$$\text{aluminum hydroxide} \rightarrow \text{``aluminous'' acid}$$
$$Al(OH)_3 \qquad\qquad H_3AlO_3$$

It is possible to follow these processes by adding some litmus to the solution. That produces another effect, which is important in the dyeing industry. As soon as a precipitate of hydroxide is formed, the solution loses all or nearly all of its color, because the color joins the precipitate. We can even say that it is predestined to retain color. By itself it is colorless and fairly transparent. As a result it is not very conspicuous (in contrast to the glaringly white calcium precipitates). We can use this phenomenon technically to fix plant pigments to fabrics. This process is called mordanting. It is chiefly aluminum sulfate (alum) that is used as a mordant. The normal hydrolysis is already sufficient to achieve this.

These last two processes also show the characteristic of oscillating between two states followed by a fixation into a third state. The first example of this was aluminum oxide.

Aluminum in the Crust of the Earth: Clay

When we analyze the crust of the earth, we find that aluminum has a prominent position among the products of analysis. It comes third after silicon and oxygen (O 49.4%, Si 25.8%, and Al 7.5%).[64] Along with calcium and silicon, it belongs to the most important crust-forming elements. In their chemistry and in the crust of the earth, both calcium and silicon represent a one-sidedness in which they oppose each other. Aluminum more or less takes on a middle position: it can act as a base, like calcium, or as an acid, like silicon. In rock-forming minerals, it often is the center of complex structures. For example, in the case of feldspar it stands between an alkaline (Na_2O) and an acidic component (SiO_2).

When granite or related minerals weather and disintegrate, aluminum compounds take a path which is again a middle road between that taken by the two polar classes of substances. The alkaline constituents are taken away by rainwater in solution, and the silicic acid is mainly pulverized and is left behind as sand. Aluminum silicate is carried away in the form of minute particles floating in water, and in the end, settles somewhere as clay.

It is possibly one of the most characteristic properties of aluminum, that it hardly circulates in natural waters, despite the fact that it can form very soluble compounds.

Good clay soils are considered to be the most fertile soils. They have the advantageous property that they not only retain water strongly, but also retain many inorganic and organic substances, even when these are very soluble. Since these substances are hardly washed away by rainwater, plants are able to profit abundantly. This remarkable behavior of clay is comparable to that of aluminum hydroxide in relation to the plant pigments.

Although silicic acid and lime belong to the substances which are taken up and processed by plants in appreciable amounts, generally this happens only slightly with aluminum compounds. Here also, aluminum holds fast to its role of mediator. The aluminum compound itself is not assimilated but rather it holds fast those substances which should be assimilated, and makes them readily available.

With the elemental circle as our reference point, it can stimulate us ever again to look and acquire new insights by comparing the substances among themselves. In this case there is reason to compare the substances which are found in the clay-like soils with those found in the atmosphere. The atmosphere exhibits a great transparency for light, heat radiation and the like. Clay is very opaque to such radiation. In the atmosphere every substance has the tendency to expand into the most rarefied, most mobile state. Air lets go of water vapor without any resistance, after it has condensed to water. Clay absorbs water and retains it, so that very little movement is possible. Other substances are also taken up by water and held back.

If we treat clay in a certain way, we produce a very plastic, formative, potter's clay. Here aluminum's "character" also becomes very apparent: to be open to influences, to be determined from the outside. When the clay has dried, it becomes hard, and the impressed forms remain. If the clay is now also strongly heated, then it achieves a very permanent, stone-like state, as shown with tiles and earthenware. With that its ability to be influenced has gone; however, the imprint of previous influences is retained.

Alchemical and Mythological views of Pottery and Porcelain Manufacture

Curious, profound laws lie at the heart of the production of earthenware, and even more so with porcelain. These only become clear when we take note of certain traditions. The first tradition is that of the four elements. This is extensively revealed in brickworks and potteries. The most important raw material is earth. With the aid of water this is made plastic. The manufactured pieces are exposed to air to dry. The process comes to a conclusion through the usage of fire in the firing.

The second tradition we find in the many stories about the creation of the human being. Very significant is the myth of Prometheus. Focusing on the technical aspects, Prometheus kneaded the human form out of moist clay. Then Pallas Athene blew her breath in and thus gifted it with life and mobility. Finally, Prometheus stole fire from the heavens and gave it to the human beings, so that their spirit would be awakened. Here, also, we find a path through the four elements in the treatment of clay. But the end result is the human being endowed with spirit. From this we may conclude that pottery is an unusual imitation of the incarnation of the human being. So, it really is not strange that when we make tools or machines, we are imitating some aspect of the human organism. In some way or other this is nearly always the case, even when it is done mostly unconsciously. But ancient amphorae and tall containers show that potters imitated the whole human form. In this context it is very interesting that in warm climates, water was kept in unglazed containers to keep it cool. The water seeped through the porous walls, evaporated, and thus kept the container cool. In this way sweating was imitated.

An important indication of the relationship between pottery and the incarnation of the human being is that when baking bread the same path is followed through the elements. Only the bread does not dry in the air, but the air element enters the bread during the action of rising the dough. We saw previously that the true recipe for bread-making is based on the constitution of the human body.[65] There are even indications to be found which show that in baking bread, there was an attempt to realize laws which apply to the ideal human being. When baking bread, the biblical saying should be remembered: "Take and eat, this is my body."

In the production of porcelain, it seems people had searched for an ideal substance. Something like the striving to bring the earth to its highest heights must have been at the root of this. We can gather this from a special stage in the process of preparing porcelain. The raw porcelain material (China, or kaolin clay), which is a very pure, white clay, is ground with feldspar, and then mixed with water and other organic substances which can rot. Then the material is then left to rest, to increase the plasticity. The story goes that the Chinese made it into balls, placed them in underground cellars, and left them there for many decades. Through a process of rotting, the white color temporarily changes to black; yet, in the end, a very noble, white material appears. What is the basis for spending such effort and diligence on this process? In the days when the production of porcelain developed, the guidelines for technology stood in close connection with spiritual insights. In one form or another, we find in all important religions the story of the "Fall," which the human being and earthly substances are supposed to have made. The highest striving had thus to be directed to counter the consequences of this Fall. In the production of porcelain this Fall, this journey into the abyss, was consciously imitated, to be able better to celebrate the conquest of the darkness. The production of porcelain is a process which was originally of a prophetic nature. It gave the possibility to practice what one hoped to achieve: the elevation of earthly substance.

From the fact that the Chinese called the granular China-clay "the bones" and the partly transparent feldspar "the flesh" of porcelain (and the soda-mineral flux, the "blood), it follows that this search for the ideal state of substance was connected to the search for the ideal human physical substance.

We now come to a remarkable conclusion: The substances in the elemental circle are all of immediate importance for the human body and the whole of nature. However, from a material viewpoint, aluminum doesn't play the part in our body which we would expect from its position in nature. Yet, mythology brings aluminum, as clay, into a much closer relation to the human body than any other substance from the circle.

Nitrogen and Aluminum

Polarity and Harmony in Various Realms

When we compare nitrogen and aluminum, their striking contrasting behavior always becomes apparent, which however is accompanied by just as remarkable a correspondence. We always find a similar property, expressed in opposite ways.

Especially when comparing their chemical properties, we find correspondences and polarities combined. They correspond in that they both tend to fluctuate between acid and alkaline. But how different this oscillating is in both cases! In the case of aluminum compounds, it is the same substance which can react as an acid or a base. The acidic or alkaline-reacting nitrogen compounds, on the other hand, are totally different substances. Aluminum is a metal, some compounds of which may react acidic or alkaline. Nitrogen, however, is a non-metal which not only forms strongly acidic compounds but also strongly alkaline ones. There is no ambiguous oscillation here, but rather a strong deflection to *both* sides. Nitrogen forms acids through its combination with oxygen, alkalis through that with hydrogen.

In the way both substances present themselves and form their own cycles, we find this extreme contrast. Nitrogen is found above all as the most important gas in the atmosphere. To some extent we also find it in the hydrosphere as a dissolved gas, but in the crust of the earth it is found only in negligible quantities. In general, nitrogen tends to shed its compounds and return to its elemental state, i.e. to rarefy as an atmospheric gas. With respect to this tendency to act as a very mobile but chemically hardly active element, nitrogen is in strong agreement with the character of air. Here the substances are very intimately mixed because of their great mobility, but at the same time as carriers of chemical properties, they are very independent of each other. Thus, in the atmosphere, we find substances which are quite independent of each, even though they occupy the same space.

The aluminum compounds restrict themselves for the most part to the crust of the earth. Although there are a whole series of soluble aluminum compounds, we rarely find them in natural waters. We are usually unaware of this phenomenon, although it is very characteristic of aluminum and is very important for the whole of nature. Many aluminum compounds are carried along by air and water, but here we are dealing with small particles which settle out as the movement comes to rest. In that way, the clay and loess layers came about.

Furthermore, aluminum does not have any tendency to isolate itself or to put itself in opposition to other elements. On the contrary, we often find aluminum forming the center of mineral forming substances, around which other substances gather; the feldspars are striking examples. Aluminum also plays a similar role in the alum salts.

Very interesting contrasts emerge when we expose nitrogen and aluminum compounds to fire. Many nitrogen compounds will volatilize, if they are not already volatile to begin with. Some will decompose into various gases, often accompanied with flames. Some typical aluminum compounds show precisely opposite behavior; through heating they change to a stone-like or even a gem stone state. Most aluminum compounds are very hard to volatilize.

The contrasts really are to be found down to the smallest detail. While a series of nitrogen compounds are prominent oxidizing agent,[3] aluminum powder can be used as an exceptionally strong reducing agent. Above all it is used to extract metals from their oxides. It evolves so much heat that it has been used to make fire bombs. The aerial bombs are the result of the cooperation of nitrogen compounds as explosives and aluminum mixtures for the fire bombs.

Whoever has seriously reflected on the relationship between technology and nature will find the following of importance. When we are preparing explosives with the aid of nitric acid, then we impart properties to solids, which really belong to elements of air and fire. By using aluminum in the aircraft industry, we are conferring properties to air, which really belong to the elements of water and earth. During an explosion solids are driven apart and thereby made equal to gases; during flying, air develops a carrying power for heavy objects which we would normally only find in solids or fluids. As it has been made technically possible to venture into the air with the aid of aluminum, so with the explosive nitrogen compounds it has been possible to find ways burrow into the earth's crust.

The last contrast we want to discuss is a very remarkable one. We have previously discussed how important it is that air, and especially nitrogen, surrounds all living and non-living things as a negative sculpture, so to speak. The living beings are even being kept together through atmospheric pressure. In contrast, it is characteristic for aluminum that it is used in pottery clay and sculpture. Comparing aluminum and nitrogen in this respect again brings out the theme which became apparent in studying the chemical properties: a harmony, which goes hand in hand with a contrast. Both substances exhibit sculptural aspects one in a positive way which fills space, the other in a negative way which creates space.

This theme of harmony and a simultaneous polarity, which occupied us at the beginning of this chapter and comes up again at the end of it, must be seen in the strictest conjunction with the position of both substances within the elemental circle. Both substances are at the transition points between the two opposite halves of the circle. *On the left hand side, we find those substances which form strong alkalis, on the right-hand side, those which form strong acids.* The oscillation between acids and bases, which characterizes both substances, becomes clearly visible in their placement between the halves, and the many contrasts which can be discovered become clear from their opposite positions.

Something from the History of the Production of Nitrogen Compounds, and the Extraction of Aluminum by Comparison

In these cases we can only allude to the significant points, though it is tempting to consider them in more detail.

The preparation of nitrogen compounds from atmospheric nitrogen may be considered as one of the most significant technical achievements of this century. Basically, two historic developments led to this accomplishment. One is the history of the methods of fertilizing, and the other the development of explosives.

In fertilizing, we interfere decisively in the nitrogen balance of the soil. In the past it was instinctive how to do this. Stable manure or compost was used. From time to time, the land was left fallow. As soon as we started to exploit the soil for greater material productivity, problems arose which had to be consciously solved. Initially, it was possible to alleviate the shortage of manure by importing guano. The trend was started by von Liebig[66] doing systematic research into the chemical composition of the soil and the plants harvested. As an early nineteenth century chemist, his main concern was with the ratios of the elements. His work gave the impulse to use artificial fertilizers. Initially, the following rule was used: what is being taken from the soil through the harvest of plants must be put back. Sodium nitrate ("Chile") saltpeter was chiefly used to replenish nitrogen compounds. It was only towards the end of the century that it was discovered that the soil has its own nitrogen economy and can utilise atmospheric nitrogen. Biological methods were even developed to supply the soil with nitrogen compounds, using legumes, especially lupines. Using artificial fertilizers, we have achieved a vast increase in material productivity and have even brought large tracts of non-arable land into cultivation. That caused new problems, however. The natural resources of guano and saltpeter became depleted. The biological methods did not seem adequate to counter the shortage. In this state of emergency the techniques of fixing nitrogen directly from the air seemed to offer a solution.

Long before its use for artificial fertilizers, potassium nitrate was needed for the production of gunpowder. The necessary potassium nitrate could be found in many places. Another possibility was to convert calcium nitrate using potassium carbonate, as follows:

calcium nitrate + potassium carbonate → calcium carbonate + potassium nitrate

For example, to satisfy the demand for calcium nitrate, France used so-called "wall scratchers," soldiers who had the right to enter the houses and stables of the rural population and scratch the saltpeter from the walls.

Lavoisier put an end to this by devising an improved method for producing saltpeter by introducing the so called "saltpeter plantation." Urine and other rotting organic material was poured over a refuse heap. After a certain time, the saltpeter could be dissolved in water and thus extracted.

Later, when not only gardening and agriculture, but also the military and the chemical industry, became dependent on saltpeter, even these methods are no longer adequate. Warfare became dependent on the limited deposits of high quality saltpeter in the deserts of Chile.

In this situation, the discovery of the first technically feasible process of producing nitrogen compounds directly from the atmosphere (the Birkland and Eyde process) offered a valuable solution. Using high temperatures, now available through the use of the electric arc, it became possible to "burn nitrogen from the atmosphere" in large quantities, and from the resulting nitrogen oxide, produce nitric acid and calcium nitrate, now called "Norge saltpeter."

Later, F. Haber (1868 - 1934) succeeded in producing ammonia from nitrogen and hydrogen on a large scale. The ammonia produced could be converted to nitrogen oxide by oxidation, using the Ostwald process (W. Ostwald, 1853 - 1932). The technical details of this process need not concern us here, and may be found in any textbook of industrial chemistry. However, we must point out that, as always, technology had to follow natural pathways, and that people tried to follow both the direction of nitrogen-oxygen compounds as well as the nitrogen-hydrogen compounds. Ostwald's process, for example, is a great technical imitation of processes occurring in healthy soil.[67]

Nowadays the Birkland and Eyde process has become obsolete because of its great consumption of electricity. However, there is now a third method for the fixation of nitrogen using calcium carbide, resulting in the formation of calcium cyanamide ($CaCN_2$). It is left to the soil to convert this into ammonia and then into nitrates.

It would be worth the effort to place the history of the extraction of aluminum besides that of the fixation of nitrogen. We would then again come across a certain correspondences, but at the same time experience the great polarities.

In the first place, there are the purely historical parallels. In the case of the extraction of aluminum from bauxite ore, the problems are precisely opposite. In the modern nitrogen industry we must contend with how to conquer the tendency of nitrogen to remain in its elemental, gaseous state. On the other hand, in the case of the aluminum industry it has taken decades to find methods for the cheap extraction of aluminum metal, as an element, from the earthy ore.

Endnotes

[1] See introductory section on terminology, IUPAC chemical names versus common (or commercial) names of substances.

[2] See R. Steiner, Chap.XVI, "Organic Science," in *A Theory of Knowledge*, 1886, Anthroposophic Press, New York, 1968.

[3] Ernst Bindel/Arnold Blickle: "Number relationships in the world of matter and in the Earth's Development" in "Contributions to Substanzforschung,"Dornach, 1952.

[4] a) R. Gillespie, D.A. Humphries and E.A. Robinson, *Chemistry*, pg. 8, 31/832, 2nd Ed 1989, Allyn and Bacon,USA.
b) Marieb E.N., *Human Anatomy and Physiology*, pg. 30, 2nd Ed, The Benjamin/Cummings Publishing Company Inc, 1992.

[5] See Chapter IX, "The Halogens."

[6] See Chapter III, "Oxygen, Oxidation, Burning and Rusting."

[7] See Chapter III, "Oxygen, Oxidation, Burning and Rusting."

[8] Baravalle, H. von, *Numbers for Everyone*, 3rd ed. Stuttgart 1959; table 2, specific heat capacity.

[9] Hydrogen and carbon have the greatest heat of combustion of all elements.

[10] Henderson, L. "The Fitness of the Environment."

[11] Parkes, D. G., *Mellor's Inorganic Chemistry*, pg. 318ff, 1967 Longmans, London.

[12] Amorphous carbon is thought to include different varieties of vegetable and animal charcoals; these are more or less impure forms of carbon. See Baravalle, pg. 356; also N. N. Greenwood & A. Earnshaw, *Chemistry of the Elements*, 1984 Pergamon Press, pg. 303.

[13] See Sec. 7, "The Natural Cycles of Sodium."

[14] Compare Julius, *The World of Matter*-Book I, The Curriculum in Class 9"

[15] See Kolisko "Hypotheses free chemistry in the sense of Spiritual Science."

[16] In the case of carbon (C) and hydrogen (H_2) the heat of combustion is the same as the heat of formation of carbon dioxide CO_2 and water respectively; in the case of water we have the option of choosing water vapor ($H_2O(g)$) or water as liquid ($H_2O(liq.)$). In the following we choose water vapor. The heat of formation of CO_2 is - 393.5 kJ/mole CO_2 or - 8.94×10^3 kJ/kg; for $H2O(g)$ the heat of formation is - 241.8 kJ/mole H_2O or - 13.4×10^3 kJ/kg. The heat of combustion of carbon, expressed in kJ/mole, is greater than that of hydrogen. See *Revised Nuffield Advanced Science Book of Data*, table 5.3, 1985, Longman Group Ltd., London; Also ref. 41.

[17] See Sec.7, "Fire as an Image of the Sun."

[18] See Chapter 3, "Solution of Metals in Strong Acids."

[19] See ref.8, "Warmth Course," Lecture 12.

[20] Paraffin, from the Latin "parum affines," means slight affinity.

[21] See Smith, A. and Dwyer, Chap., *Chemistry about You*, p 359 Thomas Nelson Australia, 1986, 1988; ISBN: 0 1700 6255 4.

[22] We could say that a protein is not a salt, oxide or other "simple" chemical compound. Analysis does give a sequence of chemical elements, but this alone is not sufficient data to establish its "tertiary" structure. A more gentle analysis yields a series of amino acids. So, it is possible to main tain that the chemical structure of protein is not as determinate or obvious, as that of copper sulfate, for example.

[23] Steiner, R., Wegman, I., *Fundamentals of Therapy*, GA 27. See chapter XIII & XIV.

[24] See section 6, on Potassium.

[25] See Chapter 10, "Silver and Lead Chemistry."

[26] The Na^+/K^+ ratio = 27.6 in seawater (Garrels, Mackenzie and Hunt, *Chemical Cycles and the Global Environment*, William Kaufmann Inc, USA, 1975, ISBN: 0 91 3232 29 7); blood plasma Na^+/K^+ ratio = 28 (Smith & Dwyer "Chemistry about You," Thomas Nelson, Australia 1988, ISBN 0-17-006255 4) (note DR).

[27] See Chapter 10, "Quartz and Gold."

[28] Expansion coefficient of vitreous quartz is about 0.25×10^{-6}, ordinary soda-lime glasses about 100×10^{-6}, Pyrex (laboratory glass) 3×10^{-6}. See Greenwood, N.N and Earnshaw A.: *Chemistry of the Elements*, Pergamon Press 1984, ISBN 0-08-022057 6 (note DR).

[29] Steiner, R., *Agriculture*, lecture 3; GA 327.

[30] Steiner, R., Wegman, I., *Fundamentals of Therapy: An Extension of the Art of Healing through Spiritual Knowledge*, GA 27. Chapter XIV.

[31] See previous section 6, on Potassium.

[32] See Chapter III, "Dissolving Metals in Strong Acids," Chap. X.2, "Characteristics of the Metallic State."

[33] Chapter 7, "Silica in living organisms."

[34] On the 'fire in the eye' and the activity of looking out, See "Platonic sight" in *Catching the Light* by Arthur Zajonc, Bantam Books, 1997, ISBN 0-553-08985-4.

[35] Chapter 10, "Lead & Silver in Metallic State."

[36] It might be worth mentioning that silicon is a semiconductor, and can even be used to generate electricity by being exposed to the sun.

[37] See, "Carbon between Sun and Earth," Chapter 7, above.

[38] See Bowen, *Environmental Chemistry of the Elements*, Chapter 2; Academic Press 1979.

[39] As the deep ocean is more acid than the top, it seems reasonable to suppose that the rate of precipitation at the bottom may just be able to keep up with the rate of dissolving and that there is no net increase of limestone (DR).

[40] Steiner, R., *Mystery Centers*, especially Lect. 6; GA 232.

[41] The intramembraneous skull-bones form from surface inwards, while endochondreal limb-bones form from ossification centers outwards.

[42] Julius, F., *The World of Matter and the Education of Man* - Book I," The 7th Grade Curriculum.

[43] (DR: same as apatite; according to Holleman-Wiberg the chemical composition of phosphorite and apatite is identical; Earnshaw and Greenwood, *Chemistry of the Elements* indicate that apatite is the only industrial phosphorus mineral).

[44] See Chap. II, "Oxygen - Oxidation, Burning and Rusting."

[45] This experiment is called the "chemical flag" in Dutch and German; see for example *Mellor's Modern Inorganic Chemistry*, pg. 815/816 ed. 1967 Longmans.

[46] Hoffmann, K.A., Hoffmann, U.R., *Anorganische Chemie II* par II.3 Braunschweig 1949, pg. 436.

[47] These same properties that make asbestos such a good material for industrial usage are the cause of its health hazard. Since the 1960s it has been known that asbestos also causes cancer of the lungs, asbestosis and cancer of the lung- and stomach lining.

[48] The role of phosphorus in ATP (adenosine 5 tri*phosphate*) may also be seen as related. ATP is the chemical means to "fix" or exchange energy in living organisms.

[49] See Sec.7, "Sulfur in Protein."

[50] Note: Ammonium phosphates are widely used as flame retardants, as are urea phosphates. Another similarity – with a difference – between magnesium and phosphorus (DR).

[51] Note: Magnesium does not react with oxygen in dry air; however, in dampness it tarnishes quickly with an oxide film. This thin layer protects magnesium from further attack by oxygen. The oxidation reaction in damp air is very quick, but proceeds without any visible heat or light effects, while at higher temperatures, magnesium will burn with an intense white light, forming a mixture of magnesium nitride and oxide (DR).

[52] Although $Mg(OH)_2$ is a mild alkaline substance it will neutralize 1.37 times as much acid as NaOH, weight for weight, and 2.85 times as much as $NaHCO_3$. See Greenwood and Earnshaw, *Chemistry of the Elements*, pg. 132. 1984 Pergamon Press/UK.

[53] Phosphorus means "light bringer" in Greek; Lucifer means light bearer in Latin. Prometheus was the spiritual power which stole fire from the Gods and brought it to the people. He was punished by being chained to a rock and having an eagle rip out his liver every day.

[54] For further details see A.D.F. Toy, *Phosphorus in Everyday Living*, Chapter 2, ACS 2nd Ed. 1977, ISBN 0-841-20293-1.

[55] The reaction $N_2(g) + O_2(g) \rightarrow 2 NO(g)$ only takes place at high temperature and even then only proceeds very slowly. The reaction is endothermic; even at 2,400 K, the equilibrium constant is small 2.5×10^{-2}. This reaction occurs in lighting discharges during thunder storms and at high temperatures, about 2,300°C, in the cylinders of internal combustion engines. See ref 37a.

[56] See Section 7, "Sulfur in Protein."

[57] Not only because nitrogen is the most abundant gas in the atmosphere. The atmosphere yields: nitrogen, oxygen, carbon dioxide (C,O), water vapor (H,O), sulfur dioxide and a host of other substances in minute amounts; see for also ref. 70, table 1.2, pg. 6.

[58] See for example E.W. Russel, *Soil conditions and Plant Growth*, pg. 352ff Longmans 1973(1980), ISBN 0-582-44048-3

[59] See Chapter 7, "Phosphorus as Element," the phosphorous flame as a salt activity in the realm of sulfur.

[60] Nowadays virtually all nitric acid is produced by oxidation of ammonia; ammonia is prepared by the catalytic reduction of nitrogen with hydrogen (Haber-Bosch process), and catalytic oxidation of ammonia (Ostwald process). See Kirk-Othmer, *Encyclopedia of Chemical Technology*, 3rd Ed., Vol 15, Pgs. 853-871, John Wiley & Sons, 1981.

[61] See Greenwood and Earnshaw, *Chemistry of the Elements*, p. 509, 1984 Pergamon Press/UK.

[62] The term "Ammonium-bases" probably used to indicate that these compounds may be seen as derivatives of ammonia. However, the IUPAC systematic name of these compounds is amines $R-NH_2$, where R is an alkyl radical. See for example F.L.F. Fieser and M Fieser, *Organic Chemistry*, 3rd Ed., pg. 224ff, 1956, Reinhold Publ. Corporation.

[63] Amphoteric, from the Greek amphoteros = each (both) of two. *The Collins Concise Dictionary*, 2nd Ed., 1988, Collins.

[64] See R. Gillespie, DA Humphries and EA Robinson, *Chemistry*, p. 102, 2nd Ed., 1989, Allyn and Bacon/USA, ISBN 0-205-12933-1.

[65] See F. H. Julius, *The World of Matter*, Book I, 8th Grade Curriculum.

[66] Justus von Liebig (1803 - 1873) in later life retreated somewhat from his original extreme position. See his *Chemical Letters*, especially # 23.

[67] See E.W. Russel, *Soil Conditions and Plant Growth*, Chap.16, p 327, Longmans, 1973 (1980), ISBN 0-582-44048-3.

VIII

The Structure of the Element Circle: An Image of the Order of Nature

The Circle of Elements and the Stage of Life

Previously, we saw that every substance in the circle of elements represents its own particular domain of nature, and that all these parts together form the whole of our natural environment. That may remind us of the story which tells us that the world came about through the disintegrating of a gigantic human form. However, the tradition whereby the world can be called the macrocosm and the human being the microcosm finds therein an exact foundation. The human being carries within itself a harmonious summation of the whole world environment. At one place, this relationship between universe and the human being is particularly clear. The salt composition of blood—despite some characteristic differences which also exist—shows a marked resemblance to the composition of the salts in the sea,[1] with respect to composition and quantities of the various salts.

The salt concentration in blood is, of course, much lower than that of sea water. In both cases we find sodium, potassium, magnesium, and calcium as the elements which form the essential basis of the salt mixture.

Balanced Opposites in the Circle of Elements

We have described and characterized the different substances from the circle of elements. In the case of every substance it was possible to relate its properties with its main task, within a certain domain of nature. It becomes ever clearer, as we gather more material in this area, that we must even deduce the properties of the substance in question from its main role in nature, if we really want to understand it. In the case of such characterizations, we repeatedly started by comparing two substances which stood opposite each other in the circle. In so doing, it becomes clear that there exists a sharply defined contrast between the functions, and, thus, also between the properties of each pair of substances.

It pays to take particular note of this, as every pair of polar opposites again may be distinguished from every other pair in style and in character. We are, as it were, taught to think through the grand ideas that lie at the heart of nature. Every principle, such as polar contrasts, does not appear rigid, but in numerous variations. And to the degree that we make such fundamental ideas our own, we penetrate ever deeper into the phenomena. We can even say: the better we translate the language of nature, the more phenomena we become aware of. We observe things and find ways which previously we disregarded. We learn from nature itself what a polarity really means; we learn to recognize the ordering principles in the world and learn how to use them in our thinking.

It then becomes more clear that every substance from the circle of elements represents a specific, characteristic one-sidedness, as expressed by its properties. And all these one-sidednesses can only be encompassed in the whole world, because each one of them is precisely compensated for by the corresponding one-sidedness of a substance on the opposite side of our circle. Only through this wisdom-filled arrangement is it possible for the power of life to create an "authentic wholeness," to tame and bring the diverging tendencies of these elements into harmony.

Here we encounter a harmonious balance again, but in a very different manner than previously. Not only are there *quantitative* relations between the elements in their chemical reactions, but also very definite *qualitative* ones. Also, in this area of quality, a strict equilibrium reigns. Here, also, everything is ordered according to a principle for which the balance is the most significant symbol.

It is self evident that these relations, which are based on opposites, are not the only relations between the qualities of the various substances. The relationships also form a sort of network and web, and just as in the case with the quantitative relationships, these also form a sort of web of interrelations.

We thus learn that the universe, right into its substances, is a mighty, harmoniously-formed whole, with the human being as its center. During a human being's life, the powers of life forming a whole overcome the tendencies of the substances in his body to split apart, otherwise he could never be a summing up of nature. It is strange that while the consciousness of modern man rests on his bodily organization, he cannot penetrate the holistic power of his own life forces with his own cognition. What he can understand and what engages his consciousness is that part of the environment in which the tendencies to split apart and the destructive forces are more manifest than the formative and constructive forces. Further, because our consciousness is active in that realm which has greatly withdrawn from life, we can freely unfold our activity. There is nothing which compels us. Down to the substances, nature has been organized in such a way that the human being can develop to become a free being.

Acid and Base in the Circle of Elements

If we divide the circle in two halves by drawing a line from nitrogen to aluminum, then both halves show a clear, regular structure. In the left part, we find four alkaline-producing substances below oxygen; in the right part, we find four acid-producing substances below hydrogen. When nitrogen and oxygen combine, nitrogen starts on the path to becoming an acid-producing substance, and, surprisingly, starts to becoming an alkaline-producing substance when it reacts with hydrogen. The aluminum compounds alternate between acid and base.

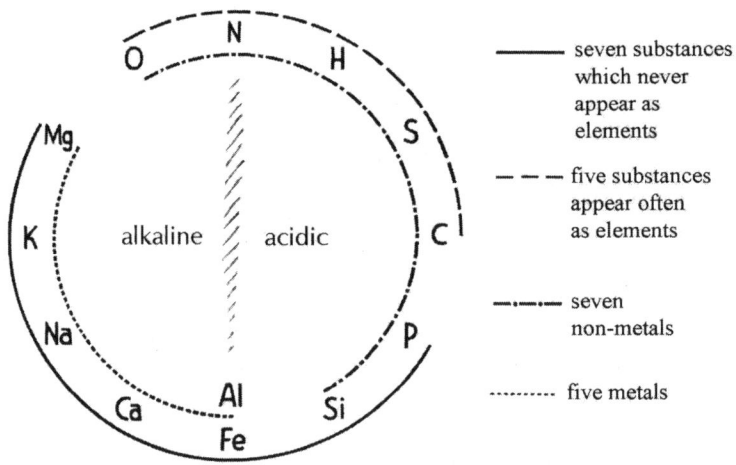

Surprising Number Relations; Air and Protein

In the circle of elements we find five metals in the bottom left half, and seven non-metals in the top right half. In the right part of the circle, we find five substances which often occur as an element in nature, and seven substances which never occur in elemental form at the left.

It is a particularly remarkable fact that just those substances which occur in elemental form—hydrogen, oxygen, nitrogen, carbon, and sulfur—are also the main substances found upon analysis of proteins.[2] With the exception of sulfur, they are also the substances which occur as the major atmospheric gasses: hydrogen, nitrogen, oxygen and carbon dioxide.[3] In its composition the atmosphere reminds us of the analysis products of egg-white.[4]

Metamorphosis in the Circle of Elements

Relation to the Annual Movement of the Sun; Seasonal Changes and the Human Organism

In the circle of elements, we not only find pronounced opposites, but also gradual transitions and changes. This becomes all the more clear when we study the natural environmental cycles of a substance and compare them with each other. For example, we can move from aluminum to calcium and the following substances towards nitrogen and then back via hydrogen along the other side. We would then discover that for the initial elements we encounter up until nitrogen, the circulation is increasingly freed from a relationship to the earth's crust, and the following series of elements show an ever increasing bonding to the earth. In this way we get a general explanation for the solubility and volatility of these elements. If we draw a line from a point between magnesium and oxygen, to a point between phosphorus and silicon, then below this line we find those substances which are only slightly volatile and whose compounds are even less volatile; above this line we will find those substances which are themselves gaseous or volatile, or which at least form some volatile compounds.

If we want to penetrate the meaning of this remarkable order of nature in yet another way, then we could use the carbon cycle as a key. This is the first conspicuous matter cycle discovered with modern methods. We find much earlier indications of a knowledge of great matter cycles in the work of the Alchemists. The characteristic of the carbon cycle is that the sun acts very strongly into the process of photosynthesis.[5] With respect to this, we have indicated[6] that in this cycle as a whole there is a clear image of the changing relationship between sun and earth during the day. Having experienced that the way processes manifest themselves in matter reflects cosmic processes, we will discover similar images in the other matter cycles, even if these are not always immediately apparent. If we compare a matter cycle which is on the whole free of the earth with one that is quite earth-bound, then we will find a similar relationship between them as was found between the sun-earth relation on a summer's day versus a winter's day. We thus come to an understanding that the circle of elements is, in itself, an image of the annually-changing relation of the sun and earth, while every substance in its own particular cycle by itself expresses the changing sun-earth constellation of a day at a specific time in the year.

I do not want to elaborate this any further. Sufficient particulars have been given in this book, which, provided we make the effort, can lead to a fruitful image.

If we make this effort, then this simultaneously implies a training. We learn to discern in what manner the imagery of the terrestrial phenomena express cosmic relations and processes. We learn ever better to follow with our thinking the hidden logic of nature. In doing this, we will have many surprising and unexpected experiences, even with everyday phenomena.

It should be clear by now that we are talking about the apparent geocentric movement of the sun – as we see it. Initially, this may seem a strange viewpoint, because we are used to always using the heliocentric system with respect to the sun-earth relationship. However, a brief reflection will show that every plant, every animal, and even every stone on the surface of the earth only experiences the sun and its movement as we observe them. Not only nature's creatures, but our own life and even large areas which are treated scientifically (meteorology, astronomy), in as far as they are based on observation, start from this naive standpoint.

The acceptance of this starting point is actually hindered, because the change in relationship between sun and earth only holds for the temperate zones. The movement of the sun and the related processes on earth exhibit a totally different course at the poles or the equator. We can avoid this difficulty if we remember that at the poles and the equator, very one-sided and totally opposite processes take place, which are harmonized in the temperate zones, so to speak. At the pole, a one-sided tendency to bring everything into a crystalline state prevails; at the equator, warmth-processes prevail, which activate substances and enhance life processes. Here we find again the contrast which is expressed by the principles of Sal and Sulfur, while in the temperate zones the Mercurial principle or the harmonizing fluctuation between Sal and Sulfur processes is more active.

The order as expressed in the circle of elements is primarily a harmonizing, balance-creating, Mercurial one. The order in the circle of elements corresponds to the annual movement of the sun in the temperate zone, because in both the Mercurial principle predominates.

We find the winter state of the earth, wherein nature rigidifies and the largest part of the path of the sun is below the horizon, expressed in the nature of those substances which predominantly form the crust of the earth and which, therefore, are rarely found in a dissolved state. We find the summer state of the earth, in which nature is in a state of exuberant activity and most of the sun's path is high above the horizon and only a very small part is below, expressed in the presence of the atmospheric gasses, which reach high above the crust of the earth.

The ascending sun movement may be found expressed in material processes such as dissolving, volatilization; we find the descending sun movement in solidification of substance, becoming denser, precipitation, crystallization.

Especially towards the end of his life, time and again Rudolf Steiner indicated that the study of the course of the seasons in conjunction with an intense first-hand experience of it, should be one of the most important foundations for any study of nature. An event which occurs in nature, be it minuscule or unbelievably large, is never a whole. It is always a part of an all-encompassing totality that only becomes visible during the year.[7] To clarify that, he points to certain processes which simultaneously take place in the human organism, while the corresponding processes in nature follow each other during the year, and thus to a certain extent are independent of each other. In relation to the processes in nature, the sequence of the seasons is a totality in the same sense as the living human form is, with respect to the processes taking place in himself or herself.

When we started on this journey to thoroughly explore the nature of the most important substances, our starting point was the human body. After having come a long way, we arrived at the sequence of the seasons. While we had to start by analyzing the human form, and thus entered into the multiplicity of nature, this variety is summarized in the sequence of the seasons, which we have come to recognize as a world-encompassing human form.

Endnotes

[1] Smith, A. and Dwyer, C, *Chemistry about You*, Chapter 21, p 359 (Table 21.1), Nelson, 1988.

[2] Instead of using the term "components" or "building blocks" (German "Aufbauproducte"), I have used the expression "substances found on analysis" which seems phenomenologically more true.

[3] Although hydrogen is not a major atmospheric gas by volume (accounts for only about 5×10^{-4} vol%; but in the tropics can reach about 4 vol%), in elemental form, it is characteristically a gas; water vapor (less than 1 vol%) and carbon dioxide are very important gases, too.

[4] The average % composition of proteins is: 53.5% carbon, 1.6% sulfur and 22.1% oxygen, 15.7% nitrogen, 7.2% hydrogen. The three gaseous elements are also important constituents of the atmosphere, but not in the same amounts.

[5] Called the "carbon dioxide assimilation process" in German.

[6] See Chapter VII, "Significance of Carbon for the Life-processes of Plants"; See also Chapter I "Man and Nature between Levity and Gravity."

[7] Steiner, R., *Mankind and the World of the Stars*, Lectures 11/26-12/31/1923, Dornach, GA 219. See especially Lecture 10, 12/29/1923.

IX

The Halogens

Fluorine, Chlorine, Bromine, Iodine - A Characterization

Proceeding to the halogens, we are now dealing with elements which obey totally different laws from those of the substances in the circle of elements. It is as if we are entering a totally different realm. That does not exclude finding substances of both groups next to each other in nature, or finding them in mutual interaction, which is very intense in the halogens. They can hardly form compounds with each other,[1] and they only occur as elements in small amounts. They are, thus, mostly found in conjunction with other substances from the circle of elements.

The properties of the halogens are so similar that we are inclined to speak of a kinship. Together they form a "family of substances." We have a greater incentive to consider them this way, because their differences show a certain regular pattern. Together the four halogens form a totality which exhibits a closed structure.

To get acquainted with these substances, we can start by showing their most important salts: fluorite (CaF_2), a beautifully formed crystal of table salt (NaCl), a few of sodium bromide (NaBr), and some potassium iodide (KI) crystals.

The first thing that becomes apparent is the cubic, or at least rectangular, form of these crystals. Because this form, so connected to the earth, is expressed so clearly, we may expect the halogens to have a prominent connection to the earth, and only a slight relation to the extra-terrestrial.

The whiteness of these crystals is even more remarkable. In the past, table salt (halite) was even seen as an example of purity. If we frequently find these crystals lightly colored, this is because of admixtures.

As a next step, we could present the halogens as elements.

Iodine at room temperature presents itself as black-brown, lustrous platelets. They give off an unpleasant smell. If we *were* to place some in the palm of our hand [DANGER], the skin there is permanently turned brown. These phenomena indicate that iodine is volatile and easily changes into a vapor, yet immediately returns to a solid earthy state and spreads out along the ground.

Bromine, in a bottle or small dish, is a dark brown, opaque liquid, above which a dark brown vapor hangs. When we want to pour the liquid [DANGEROUS], its vapor comes out first and then "pours" downwards. If we pour some bromine into water, we see small, oily drops which quickly sink down. The water above the bromine takes on a brown color due to the dissolved bromine, which gives the impression of the bromine evaporating under water.

Chlorine is a dirty, greenish-yellow gas, with an unpleasant pungent smell. If we let it escape from a container, it sinks and spreads along the ground. Because of this property and because it acts destructively on the lungs, it was used in World War I as the first poison gas.

In general, we will not be able to show fluorine, because it is difficult to prepare and just as difficult to keep it as an element. It is also a gas, but lighter and more colorless than chlorine. Most substances brought into contact with it ignite spontaneously.

Without doubt these substances form a sequence. Iodine is a solid, bromine a liquid, chlorine a vapor, and fluorine a gas which causes fire when in contact with other substances. Here we find the order of the four elements, but so modified that a special inclination to the earth becomes apparent. In the first place that is demonstrated by their vapor density. The strongly colored vapors of iodine and bromine have such large vapor densities, that, on being released, they quickly sink and spread along the ground. Chlorine, also, is nearly twice as heavy as air. Fluorine gas is the least dense, but still denser than air.

It is difficult to properly express the impressions these substances make on us. We would need to choose words which might well be seen as exaggerated or excessive. The following approach might be acceptable: Chlorine and sodium (ultimately sodium hydroxide) are produced simultaneously from sodium chloride Sodium chloride has been precipitated from the oceans in recent times and in the distant past. Imagine letting some sea water evaporate. Slowly, a white glittering crystalline mass appears. Let us now assume for a minute that we could induce sodium to weaken its capacity to bind chlorine—murky fumes would now evolve, which would kill every living creature it met, which would quickly remove colors, and finally eat its way even into more solid substances and soften them up. We ourselves would have experienced an awful smell, as if we encountered deadly, cold evil, and immediately a violent defensive convulsion would take hold of our respiratory system, coughing, and shortness of breath. If we could not escape quickly, then our lungs would soon be totally destroyed and filled with lymph (lung edema) which would result in death from suffocation.

The sensation bromine brings about is even worse, if possible; its action are just as destructive. A drop of bromine on our hand results in a deep wound. And it is not even possible to remove this drop with water. Only petroleum oil can help.

Iodine vapors are poisonous. However, we can safely sniff them without any noticeable harm (since the quantity is small). The smell is similar to that of the other halogens, but much less aggressive.

The chemical character of the halogens is seen most clearly when we consider their relationship to oxygen.

Chemically speaking, fluorine is the most active substance we know of. Chlorine and bromine still belong to the substances which chemically are very active. There are hardly any oxygen compounds of fluorine;[2] those of chlorine and bromine are unstable; only those of iodine are stable enough that they occur in nature.

The order found here is completely contrary to that usually found. Normally, the chemical activity of an element only comes to the fore when it has been oxidized. Furthermore, oxygen usually reacts more easily with an element the more reactive it or its compounds are.[3]

Thus, in the halogens we find an activity which in other elements is first induced by oxygen. We have learned to view oxygen as an emissary of the sun; particularly the awakening of activity and enhancement of the interaction between substances must be seen as an extended sun activity. The avoidance of oxygen's activity, as a result of their own extreme activity, must be seen as a repulsion of the sun impulses. That is the main characteristic of the halogens.

The halogens not only resist oxygen's influence, but they even act as its competitors and adversaries. In the case of chlorine, it is best to speak of competition, because it has about the same chemical power as oxygen. In some cases, chlorine replaces oxygen, and in other cases oxygen chlorine. Fluorine clearly acts as oxygen's adversary. Without exception it replaces oxygen in nearly all compounds and combines with the reduced substance.

Chlorine

The most important raw material for the production of chlorine is sea salt or table salt (NaCl). That implies that there is a relation between chlorine and sea water. During the production of chlorine by the old method, salt was mixed with sulfuric acid, which liberated an acrid vapor: hydrochloric acid gas. This gas is very soluble in water and reacts as a strong acid, called muriatic or hydrochloric acid. Smelling the vapor or tasting the solution hardly conveys anything specific. This substance mainly expresses "acidity," and "saltiness" as table salt. Other properties are less predominant. Of the pungent qualities of chlorine, nothing remains.

If we treat hydrochloric acid gas, or its solution, with an oxidant such as potassium permanganate ($KMnO_4$) or manganese dioxide (MnO_2), then something amazing happens: The colorless vapor becomes a dirty yellow-green, and the foul smell of chlorine appears. We have here the very unusual occurrence whereby an element is freed through the action of oxygen on a compound. Furthermore, during the transition from compound to element, the strength of the activity has *increased*, despite the fact that hydrogen chloride gas belongs to the group of strongest acid-producing substances. Chlorine is a substance which, as an element, is more active than its most active compounds. Even a giant in the area of chemical research like J. J. Berzelius (1779 - 1848) was misled by this. For a time, he considered chlorine to be an oxide.[4] He could not believe that an element could be freed in that way and could exist like that.

Nearly all elements are attacked by chlorine. Mostly a lot of heat is evolved during the reaction. In the case of many compounds we can get the impression that chlorine is being side-tracked, because with water those compounds very easily change into other more stable ones. This does not apply to the metal and hydrogen compounds of chlorine, which are stable.

A hydrogen flame, when placed into a chlorine atmosphere, will continue to burn, only we will see the flame change color. More or less the same applies to the flame of natural gas (methane). A flame of a candle will also continue to burn in chlorine, only it becomes smaller and its color changes to orange, and a dense column of soot rises. Chlorine combines with hydrogen from the hydrocarbons which make up the candle, and liberates elemental carbon as soot.

Without exception metals are attacked at room temperature by chlorine, irrespective of it being gaseous or in solution. A piece of sodium will spontaneously ignite without being heated in a chlorine atmosphere. A piece of gold-leaf, whatever its thickness, which is inactive in the presence of most other substances, quickly changes in the presence of chlorine.

Of special significance is the fact that chlorine forms proper salts with the metals it attacks ($AuCl_3$, $NaCl$). That is the reason why chlorine and its three family neighbors are called "halo-gens" or "salt formers." This name expresses the essence of these substances, because in nature we find them mainly in the form of salts.

The only elements that do not react with chlorine are fluorine and the noble gasses.[5] Moreover, chlorine does not react directly with oxygen, nitrogen, or carbon, and also not even at elevated temperatures. However, indirectly it is possible to make compounds with these substances.

We must take a closer look at the phenomena which occur due to the interaction of chlorine and hydrogen. If we burn hydrogen in a chlorine environment, we can easily follow the chemical changes with a piece of moist blue litmus paper. But in an atmosphere of pure chlorine, although a piece of litmus paper quickly loses its color, it will not change to red. By letting the hydrogen flame burn for a while, the blue litmus paper will then turn red. Evidently, an acid has been formed, hydrogen chloride gas (HCl). So, from the reaction of hydrogen and chlorine a real acid instantly results, just as a salt is the immediate result of the reaction between chlorine and a metal.

Hydrogen chloride solution is a strong acid, but without oxygen being present, and the chlorides are salts without oxygen. Water, which in the case of other acids mostly constitutes one of the components, is also absent.

This manner of acid formation is unusual. The method mentioned previously to prepare hydrochloric acid, whereby table salt (sodium chloride) was treated with sulfuric acid, corresponds in every way with the conventional way to prepare a volatile acid from their salts.

During the combustion of hydrogen in a chlorine atmosphere (or of chlorine in a hydrogen atmosphere, which is equally possible), a certain volume of chlorine will have reacted with an equal volume of hydrogen. The reaction may be represented by the following equation:

$$H_2(g) + Cl_2(g) \rightarrow 2\ HCl(g)$$

It is thus obvious to investigate what properties a mixture of equal volumes of chlorine gas and hydrogen gas have. In the dark, at room temperatures the mixture does not change. If it is ignited, then the mixture explodes, and an amount of hydrogen chloride is formed, which, when cooled, fills the same space as the original mixture did.

By allowing a mixture of chlorine and hydrogen gas to diffuse together in daylight, then a slow but steady reaction occurs whereby hydrogen chloride is formed. Direct sunlight or the light of burning magnesium ignites the mixture and causes it to explode. As soon as we bring a flask filled with such an explosive mixture of equal volumes of chlorine and hydrogen gas into bright sunlight, it explodes.

By naming the hydrogen-chlorine gas mixture "Chlorknallgas" (chlorine explosive gas), we are comparing this to "explosive gas proper", a mixture of hydrogen and oxygen gas with 2 parts hydrogen to 1 part oxygen, which burns to water. As a matter of fact, we should always compare the reactions of chlorine gas with those of oxygen gas; the combustion of hydrogen gas in chlorine with that of hydrogen gas in oxygen. In the first case, the result is an aggressive, one-sided hydrochloric acid gas, and in the second case, a mild, universal mediating substance—

water. In certain cases, such a comparison is forced upon us. We have already seen that with the aid of a strong oxidant, we can liberate chlorine gas from hydrochloric acid:

$$2\,HCl + O_2 \rightarrow H_2O + Cl_2$$

However, it is equally possible to liberate oxygen gas by reacting chlorine with water. To this end, chlorine gas must be passed into water, which now takes on a greenish color and gives off the notorious shady smell. This solution, called chlorine water, is stable in the dark. But under the influence of light, chlorine water begins to change: color and smell disappear and an acidic reaction is noticeable. In addition, oxygen gas is evolved. This experiment can be demonstrated very graphically by using a flask placed upside. If we place it in front of a window, then we will see an oxygen bubble growing at the top of the flask. We can put the whole process in the following reaction equation:

$$2\,H_2O + 2\,Cl_2(g) \rightarrow 4\,HCl(aq.) + O_2(g)$$

As we can see, in one case, chlorine gas is liberated through the action of oxygen, and in the other, oxygen is liberated through the action of chlorine. This experiment is the best way to demonstrate the rivalry between the two substances.

It is also striking that, just as is the case with carbon dioxide, oxygen is liberated under influence of light, while it is already the second case where chlorine gas has disappeared under the influence of light. We can get the impression that here light is bringing about a sort of clean-up, an opposition against something very ominous. However, we must be careful with such judgements, as there are also processes where chlorine gas is liberated under the influence of light (decomposition of silver chloride), as we shall see further on.

A very significant characteristic of chlorine, relating to the contrast with oxygen, is the manner in which both substances produce acids, bases, and salts. In the case of chlorine, everything is simple. If it comes into contact with a metal, then it rushes towards the ultimate goal, the formation of salt. In a single jolt the great transformation is made from the malleable, opaque, shiny metallic state to that of the precisely-formed, transparent crystalline one. With hydrogen gas, chlorine immediately forms a strong acid.

If we take all of this into account, then it seems as if oxygen is following some extensive ceremony. If we want to produce copper sulfate salt from sulfur and copper, then first we must thoroughly burn sulfur and pass the result into water to produce sulfuric acid. Then, we must oxidize the metal to get a base, and in the end we have to treat the metal oxide with sulfuric acid to arrive at the salt.

In conjunction with the curriculum of grade 10, we discussed how such processes proceed according to an order in which a mirror image of the larger order of Nature is reflected. More than other chemical elements, oxygen is allied with the totality of the world and reflects this in the processes in which it takes part. Chlorine has the tendency, as have the other halogens, to keep itself to one side of the oxygen world, so that the processes in which it participates show nothing of the regularity and harmony which hold sway where oxygen is produced.

With respect to the above, it is very interesting that chlorine can accommodate itself to the laws of oxygen chemistry. There are acids of chlorine, such as chloric acid ($HClO_3$), perchloric acid ($HClO_4$) and their respective salts, the chlorates and perchlorates, which are comparable, as far as their structure goes, with the acids of sulfur and phosphorus.

Whereas it is possible to prepare these by making sulfur and phosphorus oxides first and then the acids, in the case of chlorine it is exactly the polar opposite. We cannot prepare the oxides without first preparing the acids or the salts and then extracting the oxides. For example, potassium chlorate is prepared by passing chlorine gas into a hot solution of potassium hydroxide. By pouring some concentrated sulfuric acid over a few crystals of potassium chlorate, a yellow gas ClO_2, or chlorine dioxide, evolves. This gas has the curious property of exploding at the slightest provocation without being enclosed. The easiest way we can demonstrate that is by holding a flame near the gas. Oxygen and chlorine separate with a bang or crackling.

Chloric acid ($HClO_3$) is one of the stronger acids; perchloric acid ($HClO_4$) seems to be even stronger than sulfuric acid.[6]

Furthermore, it is quite obvious that here we have substances which have strong oxidizing properties. For example, if we heat dry potassium chlorate, then it expels all of its oxygen.

$$2\ KClO_3(s) \rightarrow 2\ KCl(s) + 3\ O_2(g)$$

It is a good idea to give sufficient attention to the electrolytic preparation of chlorine, and in conjunction, to the preparation of the hypochlorites [salts of hypochlorous or chloric (I) acid], which play such a large role in bleaching powder or calcium hypochlorite [$Ca(OCl)_2$] and as a bleaching agent (sodium hypochlorite $NaOCl$). Here, we must pass these by so as not to lose ourselves in too many details.

In terms of mass, chlorine does not rate high among the elements constituting the earth's crust. However, chlorides are found everywhere. As a result of the weathering of rocks, it is washed out and as chloride in solution, taken to the sea. It thus follows the same direction as potassium and sodium. And in actual fact, in nature as in our body, we do find it mostly together with sodium as table salt (NaCl). In common with sodium, chlorine has a specific affinity with the liquid cycles, both the large ocean currents circling the earth, as well as to the smaller one of blood movement in animals and human beings.

Table salt (NaCl) is the only inorganic substance besides water and oxygen gas which we assimilate directly. This fact is also important in understanding the role of chlorine.

In the past, table salt has been regarded as *the* salt. Time and again it has been taken as the symbol for purity and stability. And indeed, even to our blood which is all in movement and where it is found in dissolved form, it gives a firm basis because of its precisely governed concentration. The purity of table salt is no doubt, therefore, of such particular importance, since it is based on a struggle, on a "keeping within bounds" of the strong dark forces of chlorine. In our stomach chlorine is half liberated by our body. There, it is in the form of hydrochloric acid. It acts in a deadly way on everything that enters the stomach, and prepares the start of the digestion. So our digestion is not only based on the peaceful basis of chlorine conquered as sodium chloride, but it also utilizes its half liberated, aggressively deadly properties.

It is characteristic of our civilization that ever larger amounts of chlorine are being produced[7] and that we use its destructive properties to an ever larger degree. Chlorine never occurs as an element in nature; it is, therefore, incapable of unfolding its deadly, devastating properties. It is built into the foundations of nature, particularly in the rocks and the salt content of water. What is thus kept under control through a wisdom-filled order in nature is used by human beings after it has been liberated from that bond. Especially symptomatic is its use as one of the first poison gases in the first World War. It is so dense that it spread along the ground and filled the trenches and the shell craters, destroying the lungs of the people in there.

Also, in the more complex poisons developed later, even more horrible in their effects, chlorine is often incorporated.[8]

Also, the insecticides, the very dangerous poisons used against insects, often contain chlorine as an essential component. Their development goes parallel with that of the poison gases.

As a consequence of the general usage of chlorine in swimming pools and as a disinfectant for tap water, most people nowadays are subjected to the continuous influence of chlorine.

In light of these facts, we must once again make the comparison with oxygen. Oxygen is the respiratory gas. The lungs are particularly attacked by chlorine. Oxygen counters all attacks by poisons. Chlorine is the main component of an array of synthetic poisons, especially respiratory poisons. There is a relationship between oxygen and chlorine as between life and death.

Fluorine

In many respects fluorine behaves as an exaggerated form of chlorine. Chlorine attacks nearly all elements, which in a whole series of cases is accompanied by fire. Moreover, either quickly or slowly it destroys living beings or their remains. Fluorine does the same, but more violently and quickly. Its chemical action on other substances is nearly always accompanied by fire and even explosions.

The few substances fluorine does not react with are: oxygen, chlorine, nitrogen, and the noble gasses.[9]

Also, in its negative aspects, it surpasses the behavior of chlorine. When fluorine comes into contact with an oxygen compound, it will tear it apart into its components, reject the oxygen, and combine with what remains.

It happens that in complex compounds, fluorine takes the place of oxygen. We should not see this, as is often done, as an indication of a kinship with oxygen but rather as a consistently pursued opposition.

It is very remarkable that fluorine will attack carbon as soot or charcoal at room temperatures and cause it to ignite.

The affinity for hydrogen is so great, that we could well call fluorine a "hydrogen hunter." If hydrogen flows into a fluorine atmosphere, it will ignite immediately and hydrogen fluoride (HF) will result. Even by lowering the temperature, it is difficult to curb fluorine's affinity for hydrogen by even a small amount.

Liquefied fluorine, cooled to -210°C, will ignite immediately in hydrogen. Bringing solid fluorine and liquid hydrogen together results in an explosion, even when the temperature is nearly absolute zero (253°C).

From such observations, it follows that fluorine has an activity of its own, only achieved by other substances at high temperatures. In the past we found that hydrogen is a substance which behaves very much like warmth. We have come to see the extreme facility and speed with which it expands as ideals of warmth, as it were. Other substances only acquire such properties at much higher temperatures. In hydrogen we have a substance which behaves at normal temperatures, as others would behave at much higher temperatures. We find something similar in the case of fluorine, except now with respect to chemical properties.

Passing fluorine into water results in the immediate production of hydrogen fluoride, while oxygen is discarded. In the process oxygen is mixed with ozone and, thus, strongly activated.

If fluorine acts on hydrogen chloride, chlorine is liberated explosively.

In general, we can say that fluorine destroys nearly every compound it meets and forms new simple compounds with the liberated elements.

However, fluorine is not a substance that only has more pronounced chlorine properties. In certain instances it has its own specific properties by which it clearly distinguishes itself from the other halogens.

For example, its relation to calcium is very remarkable. Calcium chloride is very soluble in water and even very hygroscopic. We can justly regard it as one of the main salts found in the oceans. Calcium bromide and iodide are also very soluble. By contrast calcium fluoride (CaF_2) is insoluble. This means nothing more or less than that we will hardly find fluorides in the natural water, notwithstanding these always contain lime. *Fluorine avoids the water sphere.* We find it as part of some very characteristic minerals. In minerals like granite and similar we find it in low percentages as apatite [$Ca_5F(PO_4)_3$]. Apatite is fairly hard. Calcium fluoride we find as feldspar. This can form large cubic, half-transparent crystals. For the production of aluminum, cryolite or icestone (Na_3AlF_4) has become very important.

It follows from the previous remarks that with respect to its compounds, fluorine has a similar relation to the earth elements as chlorine has to the water element.

This relation to the earth element is of such a principal importance that the human and animal body use fluorine to form the hardest parts. We find fluorine in small amounts in the skeleton, but particularly in tooth enamel.

Where the human being has the strongest connection with the earth element, where the body is nearly totally mineralized, and where the most drastic mechanical actions take place, there we find fluorine.

While chlorine is mostly found in combination with sodium and always follows the circulation of the blood, fluorine is mostly found in combination with calcium, in so far as it as the tendency to congeal into a rigid shape.

The relation of fluorine to silicon and silicon minerals is very remarkable and surprising. Elemental silicon in contact with fluorine will ignite violently, while being invisible. A volatile compound has been formed, silicon fluoride (SiF_4). Two substances, which both distinguish themselves by their tendency to form minerals, form a volatile substance when they react! The volatility of silicon fluoride, together with its solubility in water, is of greatest importance in industry.

We can etch glass with hydrogen fluoride, because glass decomposes when producing silicon fluoride. If hydrogen fluoride gas passes over glass, it leaves the surface opaque. A solution of hydrogen fluoride takes some of the glass away, but leaves the surface transparent. Hydrogen fluoride also attacks other silicates.

Fluorine itself does not attack glass, as long as it is dry. The slightest amount of moisture will cause it to affect the glass. Quartz resists hydrogen fluoride quite strongly. It is only affected at temperatures above 250°C. It then transforms totally to a gas.

We must undoubtedly again see these phenomena in conjunction with the contrast that exists between the halogens and oxygen. Here, the contrast becomes absolute, in a manner of speaking. In quartz we can experience the mineral wherein oxygen is relaxed and has completely come to rest. Silicon, more than any other substance, depends on oxygen to form a mineral. In quartz the oxygen process has come to its highest expression. The result is a substance wherein the earth opens itself completely and shows itself receptive to the sun. And particularly here, fluorine acts in a most surprising way. The properties of the most noble and most durable part of the earth are transformed into their exact opposite. Form vanishes, and the power that underlies existence disappears. In a way we experience a reversal of evolution: part of the best formed during the evolution of the earth, is destroyed and transformed into a state of chaotic shapelessness, which we would expect to be the start of evolution.

When fluorine takes part in the formation of minerals, it does this in combination with aluminum and especially with calcium. This points to the fact that it has a strong connection to the earth element, which we mentioned previously but only as far as the earth, under the influence of gravity, isolates from the universe. Fluorine turns against the earth in so far as it is open to the sun.

Finally, we must mention that many fluorine compounds are very poisonous. Many are used against decomposition and fermentation.

Iodine

To begin with we must recall what was said at the beginning of this chapter[10] about the appearance of iodine.

Specifically, we know how elemental iodine occurs as very thin, black-brown colored, shiny platelets. These give off a strong, unpleasant smell, which is similar to that of chlorine but much less aggressive. It is, therefore, possible to tolerate it for a longer time. If we *were* to place some in the palm of our hand [DANGER], the skin there is permanently turned brown. Even though an iodine crystal is solid, it will not remain unchanged there but will continually spread out as a vapor. This vapor is very heavy and immediately sinks to the ground. In addition, iodine has a certain affinity for our skin, so that it directly penetrates the skin.

To study the iodine vapors more closely, we heat some iodine gently. Then, deep violet-colored vapor appears around the small dark crystals. This phenomena can be beautifully demonstrated as follows. Add a few iodine crystals to a glass flask and heat gently on all sides with a Bunsen flame. The space in the retort then becomes violet colored. However, upon cooling the color fades, and dark, star-like crystal-groups form on the walls of the flask.

Passing chlorine gas into a solution of potassium iodide results in a brown color appearing. Iodine has been liberated, while chlorine has formed a compound with potassium.[11] In general, it is a characteristic property of iodine that it can be precipitated in elemental form. Bromine also easily replaces iodide, liberating iodine. Oxidants can also liberate iodine from hydrogen iodide or iodides.

In nature we find wells with water colored brown because of dissolved iodine.

Not only is iodine easily liberated as an element, but as such it is even quite stable. That is the reason why pure hydrogen iodide can not be liberated from iodides by using concentrated sulfuric acid. The result is a purple-colored vapor, while hardly any hydrogen iodide is formed, because it is oxidized by the sulfuric acid, thus liberating iodine.

Notwithstanding this inclination of iodine to free itself from a compound, which for the other halogens would be quite stable, iodine is quite capable of combining with a large variety of substances. This is not very different from other halogens, only the accompanying reactions are generally not as energetic; fire-phenomena are hardly ever seen.

As an example, let us look at a mixture of iodine vapor and hydrogen gas. This mixture will not ignite. Increasing the temperature will result in some hydrogen iodide being formed. However, the reaction never goes to completion and, in addition, occurs relatively slowly.

We have discussed in detail how it is characteristic of the halogens to have only a slight affinity for oxygen. We had to regard as a remarkable exception the fact that there are substances, which in elemental form already show a many-sided, extensive chemical activity and yet have little affinity to oxygen. Iodine, which has the least chemical activity of the halogens, is precisely the only one whose oxygen compounds are so stable that they are found in nature. For example we find sodium iodate ($NaIO_3$) as an admixture to Chile saltpeter.

It is even possible to oxidize iodine directly using ozone or nitric acid. Using nitric acid produces iodic acid (HIO_3). That can then be transformed into diiodine pentoxide (I_2O_5), which is quite a stable compound.

Because iodine has an affinity to oxygen greater than the other halogens, it can displace them from their oxygen compounds. However, iodine itself will be displaced from its hydrogen and metal compounds by the other halogens.

Of particular interest in this regard is what Vernadsky has to say in his book on geochemistry about the natural distribution of iodine. Firstly, it has few mineral-forming properties and, thus, is rarely found as a mineral in the crust of the earth. Therefore, we find a much larger concentration of iodine in the natural waters than in the crust of the earth, in fact mainly as iodides and iodates. Many sea animals and plants accumulate and concentrate iodine. Seaweed does this to such an extent that its ash is one of the major sources of iodine. A significant amount of iodine is found finely scattered in all sorts of minerals. Because there is no strict ratio of the amount of iodine to other elements, we must assume that it occurs in a free, so called "dispersed" state.[12] It is further assumed that organisms can absorb and accumulate iodine in this form, and it can also revert to this state again. Even in the atmosphere we find some iodine, however, presumably bound to organic substances as certain spores.[13]

In comparison with the other halogens, there exists something peculiar with regard to iodine, which we can best characterize as follows. Iodine has quite a few properties which seem to suggest that it is leaving the halogen series. It exhibits a certain kinship with combustible substances such as sulfur, carbon, and phosphorus. The first of these properties is its ability to be oxidized and, as a result, its ability to form stable oxyacids and acid oxides. The second is its occurrence in nature as element, in spite of a considerable chemical activity. The third is its very slight solubility in water, combined with excellent solubility in alcohol, carbon disulfide, petrol, benzene, and similar combustible substances. If a substance is able to dissolve in another substance, this indicates a certain affinity between the two. As we have seen,[14] we can in general distinguish three such groups of affinities. On the whole, metals dissolve well in mercury, or in other metals when molten. Combustible substances often dissolve in other combustible substances; substances which have been chemically activated through reaction with oxygen, for example, are very soluble as a rule. We now see how there is a connection between iodine's weak chemical activity (in comparison with other halogens) and its limited solubility in water. In contrast, iodine's solubility in carbon disulfide shows its affinity to the non-metals sulfur and phosphorus.

Then there is also the remarkable fact that iodine is often found in the protein of living organisms.[15] That is also the case where iodine plays such an important role in the human thyroid gland. This aspect of iodine's occurrence distinguishes it from chlorine and fluorine, which appear in our organism as salts. However, this again confirms its kinship to substances like carbon, sulfur and phosphorus, which play an important part in proteins.

The fact that iodine is taken up by proteins indicates a peculiar relationship with the deeper layers of metabolism. If we ingest iodine as an iodide, then we activate our metabolism.

Iodine, however, does not only show its relationship to warmth through its similarity with the combustible substances. We have also discussed other properties which show that iodine follows warmth impulses. The first indication of this is the strong tendency of solid iodine crystals to disperse in vapor form. Even on heating, iodine doesn't change into a liquid, but passes over this stage to immediately evaporate. A second observation, which we might connect with the first, is its occurrence in finely divided form. In that sense, *iodine behaves more or less like a gas among the minerals.*

In conclusion, we must mention iodine's peculiar relation towards light and color. Using an iodine solution to filter light, we can remove some light effects of sunlight, but retain the effects of the warmth and the chemical parts of sunlight. If we produce a spectrum from light passing through ever more concentrated iodine solutions, then we observe how the visible part of the spectrum slowly disappears, while the ultraviolet and infrared part remain undiminished.

Iodine solutions exhibit exceptionally strong colors—deep brown, deep red, dark violet depending on the solvent.

Iodine compounds often have beautiful colors: mercuric iodide —bright yellow to bright red, lead iodide—the color of egg yolk, etc.

Such phenomena give the impression as if a powerful darkening effect issues from iodine. Often it destroys light; then again it calls forth colors.

Bromine

We will repeat what was said at the beginning of the chapter.[16] As far as its properties are concerned bromine stands between chlorine and iodine. In elemental form, it is liquid at room temperature, with a deep brown color which is covered by a beautiful brown vapor layer. If we pour it, the vapor comes first and quickly sinks down. If we let a jet of bromine fall on water, it will form droplets which quickly sink to the bottom, just as an oily liquid would. It forms a layer at the bottom and colors the water slightly, which gives the impression of a layer in air covered by a layer of vapor. It is more soluble in water than iodine and colors the water brown. It is much more soluble in carbon disulfide and similarly, its combustiblity is much greater in water.

The impression that *would* be made by bromine vapor on our sense of smell [TOXIC] is just frightful. In no way is it weaker than chlorine in this respect. Nor is its destructive effect on our lungs any less than that of chlorine. A drop of bromine falling on our skin immediately causes a deep, painful wound. Rinsing with water does not help; one must use petroleum solvent to wash it away.

Bromine chemistry is very similar to that of chlorine. Although the energy which is released during chemical reactions is much smaller, it is still possible for explosions or fire phenomena to occur.

If chlorine is found as chlorides in nature, it is always accompanied by bromine as bromides, in much smaller quantities, however.

Bromine also resembles iodine in the way it occurs in nature. We find it finely dispersed in elemental form in minerals. It is also accumulated in sea plants and animals. In both cases, however, it is found in larger amounts than iodine.

Bromine, like iodine, has the tendency to form compounds with beautiful colors. It is understandable that in this case it is surpassed by iodine. The renowned violet of the "purple snail" is a bromine compound.

When bromine is ingested as a bromide, it has a specific effect on our nervous system—it calms the nerves and can even be used as a sleeping-tablet.

Structural Relationships in the Halogen Group

It should be obvious from what has been said so far that the properties of the halogens show a strict relationship with respect to each other. This not only holds for their physical and chemical properties, but also for their occurrence in nature and in the human organism.

A lot of these relationships may be read from the following arrangement:

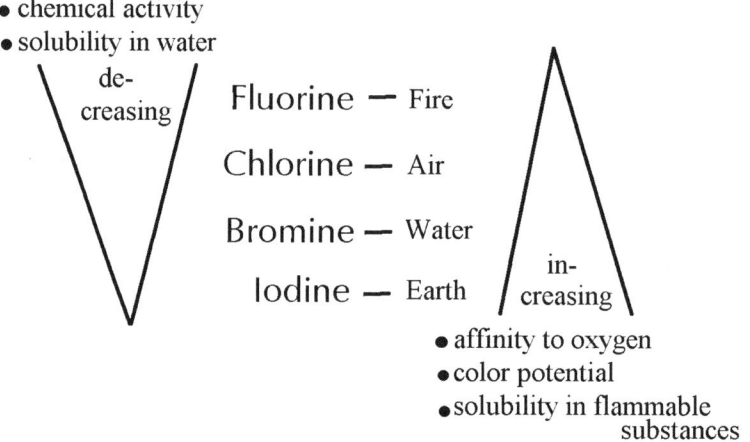

By using the properties these substances exhibit as chemical elements, a distinct relationship with the four elements becomes apparent.

When we study them as they occur in nature, then important phenomena are found which could indicate that the relationship between these substances and the elements exists. Fluorine has a particular affinity for the crust of the earth, chlorine for water. Iodine displays a certain kinship with the fire element. However, I believe that the following scheme is closer to reality.

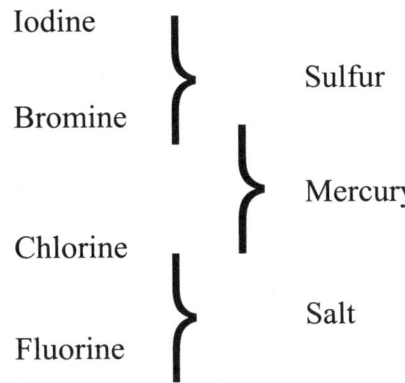

Fluorine and chlorine really only occur as salts. Chlorine and bromine are found in particular in the water sphere of the earth. Bromine and iodine have developed a certain affinity with combustible substances.

That we are dealing here with a totality or wholeness becomes apparent by realizing that in many instances, iodine and fluorine have polar properties. A few of these opposites are:

Iodine is the only halogen that produces stable compounds with oxygen; fluorine does not form such compounds.

Fluorine strongly resists being in the elemental form. In contrast iodine is easily liberated as element.

Fluorine replaces all other halogens from their compounds, especially from those with hydrogen and metals. Iodine is able to replace the other halogens in their oxygen compounds.

Fluorine predominantly forms minerals. Iodine is hardly active in this area.

When fluorine reacts with other substances, fire-like phenomena are often inevitable; iodine is the only halogen where such phenomena hardly occur.

Endnotes

[1] Since the publication of this book (1965), research into interhalogen compounds has been vastly extended. See, for example, Bailar, J.C., Eméleus, H.J., Nyholm, R.S., Trotman-Dickenson, F., *Comprehensive Inorganic Chemistry*, Vol. 2, p 1476-1563. The first interhalogen compound (IF_7) was made in 1871 by G. Gore; the last (ClF_5) in 1962 by W. Maya.

[2] Greenwood & Earnshaw (see ref #63) do speak about oxygen fluorides, p 748.

[3] See Sec. IX, "Structural relationships in the Halogen Group"

[4] Chlorine was regarded as a compound, muriatic oxide, till Davy in 1810 conclusively showed that this oxide could not be decomposed, and thus should be reagrded as a simple substance (chemical element). He gave it the name chlorine, Greek for "greenish yellow." See M. E. Weeks & Leicester, *Discovery of the Elements*, 7th Ed, 1967, p 703 ff.

[5] See endnote # 1; the first halogen-noble gas compound ($Xe^+ PtF_6^-$) was prepared in 1962 by N. Bartlett; other noble gas compounds followed soon: XeF_4, XeF_2 and KrF_2.

[6] In water there is no measurable difference between the acid strengths of sulfuric, perchloric, hydrochloric and chloric acid. As we know, the acid strength (Ka or pKa) is dependant on the solvent.

[7] Annual world production of chlorine is about 36,000,000 tons, bromine 27,000 tons, iodine 10,900 tons.

[8] Many nerve gasses (eg., sarin), poison gasses (eg. phosgene) and hebicides, insecticides and the like, are halogen compounds. These gasses have been used in many conflicts: WWI, 1936 Ethiopia, 1937-42 Japanese-Chinese war, 1963-67 Yemen, 1961-70 Vietnam, 1984 Iraq-Iran war; *Chemische Feitelijkheden* (Chemical Facts), publication of the Royal Dutch Chemical Society. See numbers 018/019 (1984).

[9] Compounds of fluorine and nitrogen do exist; the most stable one NF_3 was prepared in 1928. NBr_3 explodes at -100°C and was not prepared until 1975.

[10] See Chapter IX "Fluorine, Chlorine, Bromine, Iodine - A Characterization."

[11] It is probably more accurate to say that chlorine has displaced iodine. It is possible to demonstrate the presence of potassium chloride in the solution. The brown color is the result of iodine being liberated.

[12] However, see Goldschmidt, V.M., *Geochemistry*, p 605, edited by Muir, A., Oxford University Press, 1958.

[13] See Vernadsky's *Geochemistry*, Chap. 2-3; see also H.J.M. Bowen *Environmental Chemistry*, (ref 70).

[14] Chapter VI, "The Concept of Solubility."

[15] Iodine is found in all human tissues, but in high concentration only in the thyroid gland; see ref. 37b, p 834; [DR: the only protein which contains (or: on analysis produces) iodine is thyroglobulin which is distributed throughout the whole body; see, Mahler, H.R. and Cordes, E.H., *Biological Chemistry*, 2nd Ed 1971, p 44 table 3.1, Harper International.]

[16] See, Chapter IX "Fluorine, Chlorine, Bromine, Iodine - A Characterization."

X

The Most Important Heavy Metals

Introduction - A Comparison with the Substances of the Element-Circle

We must reach a certain rounding-off for our view of nature – as far as this can be worked out in this book – by considering the heavy metals. For various reasons it is not necessary to treat this topic as thoroughly as we have done for the other substances. First, we have available the excellent work "The Seven Metals" by Wilhelm Pelikan,[1] which can be a constant source also for teachers. Furthermore, it is not clear which metals should be discussed and in which grade. There are various grades where the metals could be discussed and demonstrated to a greater or lesser extent.

In this chapter above all we want to show how we can deduce from the most simple phenomena many aspects of the role and function of substances such as the metals in the whole of nature. We want to try to push the phenomenological method to its limit and, thus, serve anyone who seeks a methodology which will lead into the depth of nature.

Until now, we have only discussed those substances that stand out by their abundance in nature and in our body. In contrast, the heavy metals occur in only relatively small amounts, both in the crust of the earth and in our bodies. That already indicates that they have a very different significance in nature and in our bodies.

If we want to arrive at an overview of the laws and relationships which hold sway in the realm of the metals, then we cannot take twelvefoldedness as our starting point, but the number seven. We even want to restrict ourselves to the seven most important metals known since antiquity, because in them the essence of the metallic state presents itself most clearly. Each of the seven metals has very characteristic properties, which are related to certain principles active in nature and in us. However, to thoroughly explore these principles, we must proceed in a totally different way than the one we used in examining the task of the twelve important elemental substances.

An example will perhaps best illustrate the difference between the type and significance of the twelve elements from the circle of elements and those of the seven metals. Because a long tradition is associated with it, we will focus on mercury to which we have already referred several times. The Alchemists considered mercury as one of the representatives of the Tria Principia. In nature, the "mercury principle" is present especially clearly in water. There, we find a continuous change of one state into another, circulation, bringing other substances into interaction, rising, and precipitation. With mercury, we have a member of a class of substances (the metals) which especially exhibit weight, inertia, density, and in conjunction with that, an affinity to the earth. Mercury has some remarkable metallic properties: a reflecting surface, high specific density, but, in addition, a water-like mobility and volatility at normal temperatures. All the properties which enable water to perform its playful mediating role are shared by mercury. But because in mercury, the "mercurial" makes its appearance in the guise of a metal, it makes a strong impression. Therefore, it is not arbitrary that the "mercurial principle" has been given the name "mercury" or "quicksilver," but rather is something real that has been given a very exact name. It could easily appear that sodium could be characterized in the same way. We have seen how it has the tendency to bring the whole earth into a circulating liquid mass. In contrast to mercury, sodium itself participates; it leaves the metal state as quickly as possible and dissolves into water or even gathers water about itself. At the first possible opportunity, it relinquishes its separateness and loses its independence by going into the realm of water. Quicksilver occurs in nature only in small amounts. At the substance level it hardly plays a significant role. However, it brings the mercurial principle (principle of fluidity) into the metallic state, while respecting this state. It demonstrates the mercurial principle without leaving the metallic state. In keeping with this, each of the other metals are carriers of particular principles.

To a certain extent, iron forms an exception with respect to the other metals. It occurs in amounts comparable to the other elements in our element-circle. It is present in weighable amounts in our body. We can even say that iron makes its presence known in a conspicuous manner: the color of blood is determined by iron, among other things. In the crust of the earth, red and red-brown colors are mostly the result of iron. Other minerals take on a greenish color through the presence of iron. In a totally different, but no less spectacular manner, iron manifests itself when meteorites trace their luminous lines across the night sky. Iron forms a large component of meteorites. Thus, iron plays an unexpected role in the phenomena in nature. The earth and our body are substantially tinged by iron. In this regard iron stands between the element-circle and the metals. It is more active than other heavy metals in shaping specific areas in nature. It would appear as if iron has pushed itself partially into a position which should have been aluminum's, because seen from the perspective of the crust of the earth, aluminum belongs as part of the circle of elements, whereas it hardly occurs in the human body. Iron, as a heavy metal, should not be present in our body, and yet it is.

Our discussion so far indicates that we should make a clear distinction between the light metals from the circle of elements and the heavy metals. Although there are similarities, they belong to two totally different groups of substances, and, therefore, they are governed by entirely different rules and laws.

Because this has not been realized until now or has not been thought significant, we find that in textbooks the treatment of the light metals comes much too late. We need them too often in the discussion of other substances, before they themselves have been discussed. *Irrespective of whether we do or don't acknowledge the fundamental difference between the light metals and the heavy metals, the teaching of chemistry becomes much more lucid if we first discuss the light metals and later the heavy metals, as we have done in this book.*

Characterization of the Metallic State—Crystal and Metal

We must start our discussion by characterizing the metallic state as such. It appears as if we have here a sort of uniform or ceremonial dress, which many substances put on when they change into the elemental state. To arrive at a characteristic picture, we must bring together all those phenomena which occur when a metal in brought into contact with various nature forces.

Generally, in studying the nature of a substance, its relationship with light is of particular importance. There must have been a time when the substance itself was radiant. In this context, it is not so important that we think of a substance in that period of earth development as existing as a glowing gas, or whether we think of the earth as being in a more alive state in the remote past. In any case, most substances that make up the earth have made the transition from a radiant state to a dull one, to a state where inner light has left them. Many substances have retained a distinct relation with light during this transition, however, in many different ways. In certain cases, the relevant properties are so evident that we can speak of light-orientated substances. This is particularly evident in the case of the metals. When we polish a metal surface, it exhibits strong reflecting properties. In this state it will strongly reflect all light. Furthermore, metals are very opaque. Every piece of metal immediately darkens part of space. If we hold a piece of metal in our hands, we experience a strong pressure. The weight is many times greater than that of a similar sized stone. Thus, a metal strongly seeks the earth, while turning itself away from the light which comes from the cosmos. We could also say: It concentrates the earth quality in itself more strongly than do the non-metals, while enfolding at its surface a play of images of the extra-terrestrial.

Now it is also very interesting that there are transparent solids. Transparency is, of course, in the first place a property of the atmosphere. Water also exhibits this property to a high degree, although it can be lost here quite easily. Both in the atmosphere and the oceans, light presents a magnificent display. They are like palaces wherein light can live and act uninterruptedly. The crust of the earth has receded from the light sphere most of all and has surrendered most strongly to gravity. Therefore, the crust of the earth is mostly opaque. Even so, among the minerals, we find substances which are open to the play of light as soon as they are brought out into the open. These substances mostly have a precise, characteristic, crystal form.

With respect to light, metals and transparent substances simultaneously exhibit singular conformity and large discrepancies. Looking into a burnished metal surface, we see what is *in front* of it undistorted. Looking through the flat, ground, polished surfaces of a transparent substance, we see nearly undistorted what stands *behind* it, although in this case, the images have mostly shifted somewhat due to the refraction of light. In the first case we must look for the images of what stands in front, behind the metal surface.

A very conspicuous property of metals is their conduction of electricity; and they are superior to any other substance in this respect. Thus, the substances that ward off light give free access to electricity, while the transparent crystals are poor conductors. There exists this strange rule: In solids, transparency and electric conductivity never go together. Where light is free to enter, electricity is warded off or repulsed.

From this it becomes clear that not only is there a complete contrast between metals and crystals in important points, but also the above discussion shows that there is even a contrast between light and electricity.

In the so-called "Light course"[2] which was held primarily for teachers of the first Waldorf school in Stuttgart, Germany, Steiner considers the forces of nature from a very surprising point of view. He investigates the role they play with regard to our consciousness and arrives at a remarkable sequence.

When observing light and acoustic phenomena, we develop the highest degree of consciousness. Our inner capacity for imagination and mental images depends heavily on impressions of light and sound.

The fact that we are incapable of directly perceiving electricity and magnetism may be described in the following manner: These forces of nature act in a region which we cannot penetrate with our consciousness. That part of our being which is in contact with these realms, is in a state of sleep, because it belongs to the sphere of the will.[3] The phenomena of warmth we can observe directly, but our power of discernment is not nearly as finely developed as in the case of light and sound. At this level, because it is related to our feeling life, our consciousness is more dreamlike.

Thus, we arrive at the following scheme for the forces of nature:

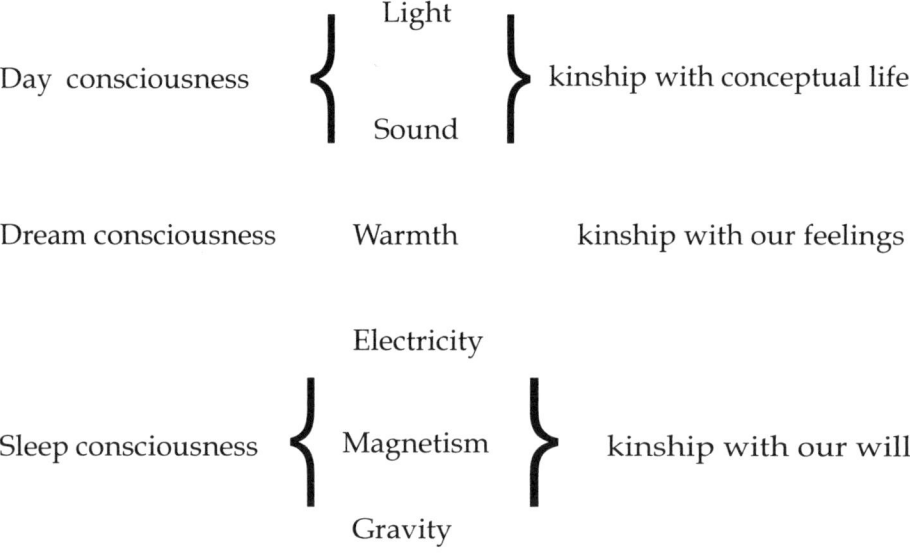

It is of great interest that warmth can work with both light and with electricity, and that this is demonstrated most clearly in the interaction with metals. Insofar as warmth travels as radiation, it is reflected by metal surfaces just as light is. Insofar as warmth travels by conduction, it is easily taken up by metals and transported further. The magnitudes of the heat and electric conductance of various substances shows a great deal of similarity.

A piece of metal is a shred of gravity; it lets in electricity, but excludes light. A transparent crystal will generally be much lighter, is open to light, but wards off electricity. Thus, crystals are open to a force of nature of which we are very conscious; metals, in contrast are open to unconscious forces of nature. There exists a certain relationship between the crystalline state of matter and our thought life, and between the metallic state, and our will. Therefore, it is no wonder that we find the best materials to make our tools among the metals.

An important difference between metal and crystal becomes apparent when we first hit a metal and then a crystal with a hammer. The metal will change shape, but remain in one piece. The crystal will not change shape in any way, but will shatter. Most metals are so pliable and ductile that they can generally be rolled and can even be drawn out. As a rule, crystals are brittle. A piece of metal is always a coherent whole, each piece of which is quickly brought into the state of the whole. (That is another expression of their good conductivity). In the case of crystals, the main point is to be found in their form, which may be seen as related to the surrounding space, rather than in their wholeness, which is maintained by matter. If it cannot maintain itself against mechanical onslaught, it will fall into pieces, which are often a repetition of the initial crystal (cleavage) form, or at least into pieces which are all the same in form. In a certain way the element of form increases.

A crystal is primarily form, but, of course, this needs to be expressed in matter to become manifest. Metals can often exhibit a crystalline structure; however, they will generally retain their typical metallic properties, such as malleability and reflectivity. Thus, a metal is primarily a substance which can bring a certain form to expression to a certain degree.

To finish off this comparison we can place a metal powder and powdered crystal next to each other. The powder of a crystal will be white, or at any rate brighter than the parent crystal; as a rule the powder of the metal will be darker. This also indicates their opposite relationship to light.

The transition of a metal to a crystal can be demonstrated in a simple manner, by adding moderately dilute sulfuric acid to zinc granules in a test tube. The metal is quickly consumed accompanied by vigorous effervescence, and soon replaced by a glasslike crystal layer.

If we hold a crystal of copper sulfate in a gas flame, then we will be able to observe for a moment pure copper. Thus, we can demonstrate the transition of crystal to metal.

Rusting as an Expression of the Nature of Metals

It is certainly not by chance that we use an acid to forge a path for light through the darkness of metal. We have discussed previously how the typical acids are all the result of combustion, a process which is characterized by being accompanied by light and warmth phenomena. This enables an acid to act as the forerunner of light.[4] In typical instances, acids have the ability to dissolve metals, and, thus, liberate them from their dark earthyness. Through the process of rusting the already earthly quality of the metal is enhanced even further. It unites with moisture, oxygen, and carbon dioxide, and condenses them into its layer of rust. Metals express their kinship with the contracting quality of earth even in their chemistry.

Metals and Planets

Tradition and Modern Research

Until now we have only studied those properties of the metals which they have in common. In reality we are dealing with very different substances, which have put on more or less similar uniforms. How can we find a way to a deeper understanding of their nature?

If we want to take the traditions of olden times seriously, then we must take into account the relation between the seven oldest metals and the seven classical planets known since antiquity.

Saturn	-	Lead
Jupiter	-	Tin
Mars	-	Iron
Sun	-	Gold
Venus	-	Copper
Mercury	-	Mercury
Moon	-	Silver

By doing this we are in danger of provoking criticism from modern science, since the direction which western scientific thinking has taken depends heavily on turning away from the old traditions. We must, therefore, proceed carefully, despite the fact that it isn't hard to point out the mirroring of the relation with the planets, even within technology. The fact that Steiner, through his spiritual-scientific research, has confirmed the traditional connection and has even contributed many new facts about the relation between the metals and planets, may not be a sufficient basis for many to base their work upon. However, following Lili Kolisko's experiments on the influence of celestial activity on earthly substances,[5] we may regard the relation between metals and planets as a scientifically proven fact. When discussing the metals, we can, therefore, safely point to this relationship in education. But even if we do not want to discuss this relationship, we can still use the possibilities which result from the ordering based upon that of the planets. We will then quickly notice that we gain a surprising instructional overview.

The Usage of Metals in Three Spheres

As we begin to discuss the properties of various metals in more detail, it will be helpful for gaining a better overview to distinguish three main areas of usage: technical, medical, and metal chemistry.

Technical Usage

We look for those locations where large amounts of the mineral have been deposited through geological causes. We extract the metal there and make use its earthy-heaviness and mechanical properties, such as its density, surface behavior, malleability, etc.

Medical Purposes

We must proceed in a totally different direction if we want to use the metals for medical purposes. We must liberate something from the metal which is rather hidden by the more or less robust properties used by industry. We use a small amount of metal or of one of its salts, preferably as formed by nature. In the subsequent process the ponderable (earthy-heaviness) aspect is eliminated, while the really essential qualities remain. In homeopathy, for example, this is achieved by a rhythmic serial dilution of a solution of the salt. We initially mix and shake 1 ml of solution with 9 ml of water, then again 1 ml of solution with 9 ml of water, etc.

Just as we can use a metal in its solid form as a tool to achieve powerful mechanical effects, so also we can prepare a metal through rhythmic dilution, so that it becomes a tool, as it were, by which precisely chosen effects may be achieved in the realm of life or even act on deeper layers of the human being. The possibility to achieve such effects demonstrates that there are processes in the human body which are related to the finer qualities of the metals.[6]

A New Realm - Metal Chemistry

A new realm is entered as we begin studying the chemistry of the metals. Any metal can be taken through many transformations. In practice, one generally strives to manufacture all those compounds which are used technically. However, just in this realm we can experience revealing phenomena.

In our more detailed discussion of the metals, we first want to focus on the properties of the metals proper, and then turn our attention to the related chemical phenomena. We will not go into a discussion of the medical aspects; we can dispense with this, as Wilhelm Pelikan has worked out such an admirable picture of this aspect of the metals in his book.[7]

Not only does every metal have general properties, but each metal also has specific ones which it alone has to that degree. By way of a thorough study of these, we discover that they form a cohesive group, a characteristic whole. We want to give these groups of properties our special attention. By comparing them with each other, they reveal themselves to be so different, that we can speak of individual "metal-characters" or even of "metal-themes." It is very interesting to note that industry had to take these into consideration extensively. Each metal, therefore, acquires its particular style. In industry we could speak of a lead style, an iron style, a copper style, etc. The comprehensive principles which are represented by the metals on earth and also by the planets in the heavens are reflected in these styles, in these "metal-themes." We have already indicated this for mercury.[8] To a certain degree, it is even possible to see a relation between the medicinal power of a metal and the style it exhibits in industry.

If we want to discuss the metals from the chemical aspect, then we must be clear that, in this area, it is often possible to portray their essence in beautiful images.

The possibilities that present themselves here have been used by L. Kolisko.[9] She assumed that diluting a metal as a salt simultaneously brought about a "setting into motion" and a cancellation of its "earthy-path." If she wanted to follow the configuration of a particular planet, then she placed a piece of filter paper in a solution of the corresponding metal salt, which was then absorbed. Or, she used a mixture of salt solutions, or she first let one solution rise and followed this with the second solution. In the filter paper powerful and expressive images were formed. The expressive gesture of these images varies as the configuration of the corresponding planet changes. It is very important with this method to let the processes occur in a plane, which greatly enhances the sensitivity. Furthermore, this forces the substance to reveal its "inner-self," as it were, and make everything visible to the eye.

We will now discuss the seven metals, using a sequence which as nearly as possible follows a cosmic order. According to Steiner the planets can be arranged in two opposing pairs.

       Saturn   Moon
       Jupiter   Mercury
       Mars   Venus
              Sun

This contrast is clearly revealed in the corresponding metals:[10]

       Lead   Silver
       Tin   Mercury
       Iron   Copper
             Gold

Discussion of Lead and Silver

Lead and Silver in the Metallic State

Lead minerals often contain so much silver that their extraction is economically possible. The metal is first extracted from the ore. Exposing the molten metal to the atmosphere causes a layer of matte oxide to form immediately on the surface. If we remove this layer, the lead underneath shines like a mass of mercury. However, the surface quickly tarnishes and becomes dull again. If we continuously blow the dull surface away, the mass of metal gradually diminishes, until there remains a residue with a very beautiful luster, which does not turn dull. Here we have pure silver. We have also demonstrated something of the contrast between lead and silver. Lead has the strong tendency to combine with oxygen and, thereby, to be converted into an earthy state. In contrast, even in the molten state, silver resists being converted into the earthly state. It forcefully maintains the state in which it can develop the greatest possible interaction with light.

If we placed two polished spheres outside, one with a silver coating and the other with a lead coating, the silver surface will show a clear (albeit spherically distorted image) of the surroundings. By carrying the sphere around we will see ever new scenes appearing and disappearing. More than any other substance, silver has the tendency to react to the activity of light in the surroundings, and to accompany these with images. In an almost ideal form, it shows a property common to most other metals.

At first, the lead sphere will also produce similar images, but less sharp and with slightly subdued colors because of its blue sheen. However, after a short while it will seem as if a veil is being drawn in front of the images, which slowly turns into an opaque gray-blue curtain. If we carried *this* sphere around, it would remain a gray-blue sphere. Images have been wiped out; the reaction to light from the surroundings is subdued.

By this we have found one of the major properties of silver: it has the tendency to hide itself by presenting images of the surroundings. The purer the silver, the more it acts as silver, and the less you see it. In contrast, lead hides behind a gray crust. Perhaps it is more correct to say: lead lets its specific property to form gray layers strongly predominate, and, therefore, its property to react to the surroundings[11] is thrust aside.

If we look at the properties of silver more closely, we will in the first place find such properties which, in a manner of speaking, aim to enhance the reflective power to the utmost. We can mention:

1. Silver has little color of its own;
2. Silver can be more easily burnished than other substances;
3. The burnished surface reflects 95% of the light;
4. Silver is unreactive with respect to most chemicals.

A second group of properties which doesn't initially seem related to the above, is the following:

1. Of all substances silver has the greatest electric conductivity;
2. Silver has the greatest thermal conductivity;
3. Silver is heavy, denser than iron or copper (its specific gravity = 10.5 g/cm^3);
4. Silver is opaque, as are other metals;
5. Silver is the easiest to roll, and to hammer, with the exception of gold.

We can bring these two groups of properties into relation to each other by considering what a silver sphere does in a light filled space. Not only does light not penetrate to the interior of the sphere, but it is reflected back more strongly than by any other substance, while the metal itself hardly heats up. Silver creates a near perfect hole in the light filled space. And just in this hole, gravity rules, while electricity has free entry more than in other cases.

Silver is the best conductor of heat, but its surface reflects heat radiation very strongly. As with all metals, but to a greater degree, silver has a twofold relation to warmth. Heat radiating through space like light is reflected more strongly than by any other substance; conducted heat, which can be sensed by touching a body, is more easily absorbed and distributed through the body than by any other substance.

In connection with this is the fact that it is very hard to make silver radiate light *and* heat. For example, we can heat various substances in the dark and observe the temperatures at which they begin to glow. Due to its black color, carbon will radiate first, while silver will remain dark to the very end. If we hold our hand next to a hot piece of silver, we will hardly feel any warmth. If we draw black letters using graphite on a piece of silver and then heat it in a dark room, we will see the letters irradiating ever more strongly against the black background. *With respect to reflection and radiation of heat and light, silver and carbon behave as complete opposites.*[12]

Silver can very easily be rolled and hammered cold to an incredible extent. Through hammering, we can even produce very thin (0.0027 mm thick) sheets of silver leaf, which appear translucent blue when held up to the light. A mere gram of silver will produce a wire thread of 2 km! In that respect it outdoes all other metals, except gold. Silver not only has the tendency to extend its surface, but also the corresponding capacity to retain its cohesiveness, and even when it is nearly all surface as with foil.

From these phenomena we see that with its whole being, silver is placed in the great polarity of light and gravity, about which we have spoken often. What applies more or less to every metal applies especially to silver. There are metals where the separation between light and gravity is quickly blurred. In the case of silver, these realms are most clearly separated. The surface of silver, for all its delicacy and beauty, has developed the property of being rather impervious to influences from various other substances, and of being especially opaque to the radiant forces of nature. In many ways, silver is a perfect metal; its essence corresponds more than for other metals, to the "metal cloak," or costume, the principle which it carries so well.[13]

Of course, the most important properties of silver are used in practical applications. The use of silver in mirrors is probably the most striking and well-known. We need silver primarily to see ourselves, or to see an image of part of the surroundings which otherwise would remain invisible to us, as in the case of a rear-view mirror in a car. The largest and purest mirrors are those found in reflecting telescopes, where a very thin layer of silver (sometimes aluminum) is deposited on a slightly concave ground glass disk. These are used mainly to make an image of those parts of the universe which would otherwise remain invisible.

If we turn such a telescope mirror *towards the sun*, then we find a real image of the sun in the focal point. The image of the sun floats above a silver bowl! If we bring an object in the vicinity of this image, powerful heat phenomena will occur, which can lead to very high temperatures. Silver is the most suitable substance to catch and direct the sun's influence. It is important to note that for manufacturing such mirrors, we really only need to apply a very thin layer of silver. We are dealing with an effect that is produced by the exterior of the substance. We can, therefore, achieve great effects in the realm of light with minimal amounts of silver. By directing the telescope to the *night* sky, we get an image of the stars in the focal plane.

In the case of the telescope we again meet one of these highlights of technology, to which we have often pointed. In this case, the properties of silver have been used to the utmost of their possibilities. At this level, images of the universe, of the sun, and of the stars are caught and directed.

Although it has an electrical conductivity, silver is only used in very special cases, because it is too expensive. However, these particular uses are quite telling. Household fuses used to contain a small piece of silver wire, which is part of the electrical circuit. If the electrical current becomes too large, this piece of wire melts and the circuit is broken.

The silversmith makes use of the malleability in conjunction with the beautiful surface of silver. We have this art to thank for producing the most delicate objects.

Silver is also very suitable for making cutlery such as forks and spoons, because its surface is rather chemically inert (except for sulfur-containing foods, such as eggs).

For the most part, silver objects are variations of the bowl form: vases, beakers, jugs, spoons, etc. What is the essence of a bowl? It is carried or rests on the ground and is open to the heavens. The gesture of this form corresponds to the being of silver.

Silver not only retains its brilliance with great force, but has something which we might call aggressive purity. Flowers last longer in silver vases than in vases made of any other material. It is possible to disinfect water using a silver sieve. Therefore, silver is also one of the most suitable substances to be used in our body to replace parts of the skeleton.

In contrast, lead is everything other than an ideal metal. All those properties which distinguish silver are possessed by lead only to a slight degree. Its burnished surface only reflects 61% of the light. Furthermore, it quickly loses its luster. Lead can only withstand a low tension without losing its cohesiveness. Its electrical conductance is the lowest of the seven metals, with the exception of mercury, which due to its liquid state cannot be compared to lead. The conductance of lead is about 1/12th that of silver (see again Pelikan, endnote #1).

Lead also has the smallest conductivity for heat, with the exception of mercury. It is 1/13.5th that of silver. It follows that the ability to retain its cohesiveness—in general a characteristic of metals—is least developed in lead.

In only one property is lead superior to silver: its specific density is larger (11.34 g/cm^3).

However, interestingly enough, some of the "weaknesses" of lead are very important in industry. This applies especially to the fact that its surface is very susceptible to chemical action, which in itself is, of course, not useful. Although lead itself is acted upon, products form which are resistant to further chemical action. Nearly all influences are brought to a standstill by lead, and, therefore, it has a delaying, retarding influence, even for the most violent activities.

For example, although lead oxidizes quickly, the layer of oxide is impervious and protects it. Therefore, lead was used for pipes, roofs, or in stained glass joints.

Lead will also cover itself with this impenetrable layer in soil or even in sea water. Lead pipes could be left in the soil without any protection. Lead is used as a protective layer for underwater cables. The use of lead in coffins when protection from the environment is necessary is well-known. Even some very active chemicals, such as sulfuric acid and hydrofluoric acid, can be kept in leaden containers. Their activity is also delayed by the resulting reaction product.

Lead is so soft that it can be scratched by a fingernail. Therefore, it was used in the past to write on paper and even on slate. It has no elasticity. Throwing a piece of lead violently against a wall will cause it to just drop, more or less flattened. It yields to every push and pull, to every twist and turn. A piece of lead that we want to use as a covering can be fashioned readily into any shape. To be transported, lead pipes are rolled onto great drums, and then unrolled on delivery. If a house settles into the soil, lead pipes will yield, while iron or stone piping will break.

Lead plays a special role in testing explosives. To that end a leaden cube is used with a cylindrical hollow which reaches down towards the middle of the cube. If an explosion occurs in it, then its shape will change in a more or less drastic way; however, no other damage will have occurred. If we had done the same experiment in a cube made from stainless steel, then we would have seen total destruction. From this we can see that in the area of mechanical and chemical influences, lead has the same delaying effect. The violent movement caused by the explosion spends itself in the change of shape which the explosion itself has brought about. The degree of change is a measure for the force of the explosion.

Tools used to grasp something in the surroundings or to reshape something else are generally made of iron. Lead is absolutely unsuitable for this. It can only undergo influences. Each effect on lead leaves as it were, an imprint behind and, thus, to a certain degree is captured. In a way, lead records the outer influences.

Lead can be quite easily liquefied through heating (melting point = 327°C) or pressure, so it can be cast in various shapes. The leaden sheath of an electrical cable is installed by using a press. Lead piping can be made by forcing lead through a sort of funnel with a center piece in the opening.

In buckshot foundries, a molten lead alloy is passed through a sieve and dropped into water. If the distance through which the lead falls is fairly large, then the drops have time to become spheres which harden in the air.

The use of lead in printing for making type is again because it can be easily melted and easily sets into a shape. Both for making buckshot and printing type, alloys are used which are harder than pure lead.

This melting and casting of lead can easily be demonstrated. If we proceed very carefully, we can even melt a piece of lead in a common test tube. It is a beautiful sight to see a brilliant mercury-like drop well up out of the gray mass of tarnished lead granules. However, it is better to use a heat resistant quartz reaction tube. We can then use more lead and pour it from a certain height onto a stone table or a piece of iron. On reaching the surface it forms strangely formed shiny surfaces. A very dramatic event takes place in front of our eyes when we heat a bit larger quantity of lead in an iron ladle and then pour it into a large container filled with water. The sudden cessation of swift movement is deeply impressive. Mostly, the mass remains more or less a whole, even when it sets in such amazingly strange forms. Watching this phenomena is a New Year's tradition in some places.

Lead, which belongs to the heavy metal group, is often used because of this. It is used as ballast for (fishing) nets, as weights in scuba diving, etc.

A very heavy metal is also always inert. Thus, lead is used in those cases where movement is to be forcibly maintained, such as buckshot or as a ballast in bullets. The nature of lead becomes apparent in a remarkable manner in its usage in dum-dum bullets. The bullet's punctured steel point-casing opens up when hitting a bone. Because of the violent deceleration, heat evolves. The lead used as ballast melts, disperses sideways and causes very large, severe wounds.

Lead and Silver Chemistry

Silver is a noble metal to a high degree. Its surface is not attacked by diluted acids, and not by oxygen even when heated. It will only dissolve quickly in nitric acid. If we touch our skin with a small moist silver nitrate crystal, then slowly black smudges of finely divided silver will become visible. In general, silver will be easily liberated from its compounds.

A crystal of silver nitrate dissolves quickly in distilled water without any cloudiness. On addition of some sodium hydroxide, a noble, dark brown-colored, opaque precipitate of silver oxide forms. To ensure that the process can be followed clearly, it is worth using moderately large glass beaker for this experiment. The emergence of this thick, fairly dark mass from the addition of two clear, colorless liquids comes as a surprise. If we add some tap water to the silver nitrate solution, a pale veil of silver chloride will appear.[14] Using sea water, which contains large amounts of chloride, then a thick milky white precipitate will form.

If the tap water does not contain too much chloride, we can create a delicate turbidity which can be used to make a lovely demonstration of the creation of color according to Goethe's theory of color. If we hold the container against the light, it will appear yellowish or reddish, while side-lit in front of a dark background, it will seem slightly blue. If the water contains too much chloride or has been too strongly chlorinated, it may have to be diluted with distilled water.

Exceptionally beautiful phenomena come about, if we produce the silver chloride precipitate in a large cylinder glass or large 2 l Erlenmeyer flask. We should place this in a dark room on our illuminating apparatus.[15] First, we fill the glass with domestic tap water and then add some silver nitrate solution. The unusual delicacy and gracefulness of the clouds of precipitation become more distinct if we have first gently swirled the clear tap water contents of the glass.

If we use the same set up and drop a small silver nitrate crystal into the tap water, then it leaves behind a precipitate in the form of a graceful streamer. If we place somewhat larger crystals as carefully as possible on the bottom of the flask and then bring the water in a fairly quick rotational movement, then the whirled up precipitate shows an imposing vortex form of a cyclone. We get a particularly beautiful phenomenon if we add some sodium chloride crystals to a clear solution of silver nitrate, placed on the illuminating apparatus. They form bright white streaks of precipitation.

As lead chloride is also very slightly soluble and is white, the same experiment may be done using lead nitrate. The precipitate will, however, be much less delicate, more opaque, heavy and grainy. It will also sink much faster. Such experiments clearly demonstrate how lead is simultaneously related to and also opposed to silver. Also, lead chemically exhibits a slow, heavy character.

It is possible to stir the silver chloride precipitate in water in such a way as to produce a milky-white fluid. Then, placing it in front of a window, it quickly acquires a black-purple color. Under the influence of light, silver chloride is decomposed. But, if we cover the beaker with opaque wrapping, then no change of color occurs.

It is extremely interesting that silver compounds in particular demonstrate such a sensitivity to light. Furthermore, it is remarkable that the reaction to light is darkness. If we remind ourselves that metallic silver, more than other metals, turns itself away from light, then this phenomenon starts to speak in a clear voice. This was used in the early methods of photography. In general, silver iodide or bromide are used today, and a somewhat different path is used to achieve the image. In any case, through photography (again through a technical application) something essential clearly comes to the fore.

To understand these phenomena properly, we must realize that in the realm of light we are always dealing with images. We observe things in our surroundings with our eyes as changing images. We can recreate clear images of these things in another place, for example in a mirror, or with the aid of lenses on a screen, or on the film in a camera. Of course, we not only have to deal with sharp but also with fuzzy images, which only have the color in common with the light source. The normal illumination through the sun or a light bulb must be seen as a transmission of images. Thus, a red colored lamp will give all objects in the surroundings a red cast; a white light bulb will radiate white light. By using a small lens we can create a sharp image of the bulb at any point. We can thus show that what appears as a diffuse brightness is in reality an image. What is generally seen as a beam of light or even as a bundle of rays is nothing more than a space filled with myriad miniature images of the light source, projected onto floating particles — smoke, fog, dust. As soon as this becomes clear, we realize that even the uniform darkening of silver chloride may be attributed to image forming. Just as the sharp image on the photographic plate is an answer to the sharp image which falls onto it, so the hazy dark discoloring is an answer to a blurred image.

In photography, just as in the case of mirrors, images are produced with the aid of silver. Here, also, we do not see the minute silver particles, but the overall image.

It is possible to carefully moisten a piece of filter paper with silver chloride. For example, first soak it in a sodium chloride solution, and after letting it dry, moisten with a silver nitrate solution. If we then subject it to light, it will take on a dark blue color. If we place a sharply defined object on the paper, then we will see an impression in the form of a white shape.

When we work a lot with silver compounds, we will notice that stains will often occur, which catch our eye because of their color tones and powerful play of forms. The "predisposition" to imaging is demonstrated by such phenomena.

Particularly important and stimulating for the students, and also easily performed, is a demonstration of how a silver mirror is made. The preliminary instructions are found in every good chemistry text. Starting point is a solution of silver nitrate. First, we produce a silver hydroxide precipitate by adding sodium hydroxide solution. Then we dissolve this by adding ammonia. By adding a reducing agent such as lactose or glucose, the solution slowly turns opaque and black-brown. If we gently pass a flame along the sides of the glass container we speed up the process at that particular point. As a result we get a beautiful, uniform silver layer. We can observe how the wall of the container becomes opaque and images of the surroundings become visible. If we use a small round bottom flask, then everyone can look at the bottom and see their distorted face. This experiment demonstrates the technique of mirror making, and, as we have often seen, the technology gives us simultaneously a clear picture of the nature of silver. From the fact that this experiment is so easy to execute and that the result is so beautiful, it follows that silver chemistry tends to express itself in precipitates in the form of mirrors.

It becomes really dramatic when sulfur, hydrogen sulfide, or sulfides react with silver compounds. Dark silver sulfide (Ag_2S) results.

Silverware goes black because of canals contaminated by hydrogen sulfide. Silver spoons get black stains from sulfur-containing foods such as onions, egg yolk, or mustard. If we add some silver nitrate solution to a solution of sodium sulfide (preferable—solutions of ammonium sulfide or hydrogen sulfide have a ghastly smell), we get a dirty brown-black precipitate. By choosing the correct proportions, the result gives the appearance of sewage water with excrement. This experiment is best done in front of a light background.

On a small scale in a *sealed flask*, we can prepare some hydrogen sulfide by adding hydrochloric acid to iron sulfide, and then hang a silver coin and a piece of filter paper moistened with silver nitrate in the vapors; we then also observe a discoloration. The experiment with filter paper can also be done on a larger scale, to achieve a more spectacular result.

Silver and silver compounds often make an especially pure and noble impression and are used accordingly. But it is just as surprising and typical for the crude natural processes, that by using this substance it is also possible to produce very nasty images of pollution and filth. At any rate, the interaction of silver and sulfur indicates that the physiological-therapeutic effects of silver penetrate in the same region as does sulfur, namely into the depth of the metabolism.[16]

Lead Chemistry

We turn now to the chemistry of lead. We have already established that lead is easily attacked. If exposed to the atmosphere it will quickly rust. But because the rust forms a protective layer, the metal underneath will not undergo further attack.

There exists a striking interaction between lead and oxygen, which expresses itself in the first place in that there are three lead oxides. The yellow-orange lead (II) oxide or "litharge" (PbO), the bright red lead oxide or "red lead" (Pb_3O_4), and the dark brown lead (IV) oxide (PbO_2). It is remarkable that as the oxygen content increases, the color deepens.

Orange litharge is produced if lead is heated under specific conditions and a powerful air supply. It is used especially as a starting point for other lead compounds; another yellow form is often used under the name of "massicott" as a pigment. Litharge will dissolve both in acetic and in nitric acid.

At this stage it is probably appropriate to mention that lead as litharge is often used in glass (crystal, cut-glass) or glasslike substances. Lead gives a greater power to refract light to glass—here again an effect which is opposed to the reflecting power of silver—and lead decreases the hardness.

To extract lead from litharge is an easy, beautiful experiment. We put a small heap of litharge on a plate. Then we place a bright flame of a Bunsen burner nearby, and with the aid of a blowpipe (or pipette), direct the flame onto the lead oxide. The mass starts to glow and soon mercury-like droplets will appear.

Red lead (Pb_3O_4) is produced by heating lead under somewhat different conditions and at a lower temperature than in the case of the production of litharge. It is often used mixed with linseed-oil as a protective coating against rusting for iron constructions (ingredient in Rustoleum). It causes the linseed-oil to become viscous and form a continuous layer. It also galvanically renders the iron passive. As we see, this "red lead" is a good representative of lead metal.

We can make brown lead (IV) oxide by pouring nitric acid over red lead. Some of the oxide dissolves as lead (II) nitrate, and some remains undissolved as brown lead (IV) oxide. This experiment is easily done. In this reaction, red lead behaves chemically as if it were a mixture of two oxides: $2 PbO$ and PbO_2. Because lead (IV) oxide has unmistakable acidic properties, we may regard red lead as a lead salt derived from a lead acid. Red lead (IV) oxide has strong oxidizing properties.

Only a few substances can dissolve lead. Best of all are dilute nitric and acetic acid. The most important soluble salts are lead (II) nitrates and the acetates.

Because of its sweet taste and white color, lead acetate is also known by the name "sugar of lead." It has often been misused to produce chronic poisoning—"inheritance powder." If we add some sodium hydroxide solution to a lead (II) nitrate solution, then a white, jelly-like precipitate of lead hydroxide results. This dissolves by adding more sodium hydroxide solution. It thus behaves as an amphoteric substance: with respect to acids it reacts alkaline; with bases it is acidic.[17] Litharge will also dissolve in sodium hydroxide solution. Here again we encounter the fact that lead is easily influenced by external factors. (Brown lead (IV) oxide is also capable of showing amphoteric behavior. Mostly, however, it reacts as a weak acid).

In the chemistry of lead, we regularly encounter thick, opaque precipitates exhibiting bright colors. Precipitates are to be expected, as the substance thereby again goes in the direction of contraction and rigidity. The bright colors come as a surprise, as they seem to be in contrast with the dull gray color of metallic lead. One of the most beautiful precipitates is made by adding potassium iodide to a solution of lead (II) nitrate. Two colorless liquids together produce a warm-yellow mass. If the concentrations are not too great, by heating we can change the mass into a clear liquid. On cooling lead (II) iodide precipitates, but now as glistening small crystals. We have here an eloquent picture of the back and forth movement of lead between mobility and rigidity under the influence of heat. A precipitate of lead (II) chloride will also dissolve on heating, but this is a less beautiful experiment. A very beautiful precipitate also comes about by adding potassium chromate (K_2CrO_4) to lead (II) nitrate solution. The lead (II) chromate that results is used as pigment under the name of chrome yellow. In another case, by adding solutions of lead (II) nitrate and soda, we get a brilliant white precipitate of lead (II) carbonate. The very important pigment "white lead" is alkaline lead (II) carbonate, a mixture of lead (II) carbonate and lead (II) hydroxide.

The three lead compounds which are used as pigments—red lead, chrome yellow, and white lead—stand out because of their brilliance and covering power, which completely fits the character of lead. They are particularly suited for use in oil paints.

Of particular interest is the old Dutch method of producing white lead. Spiral rolls of lead sheet were placed over a solution of acetic acid in loosely covered earthenware pots, which stood in rows in a layer of horse manure. After about two months, white lead is produced. It is possible we are dealing here with an alchemical process: the ugly, all-too earthy lead is buried, as it were, in a region where it has to undergo certain processes of decay and arises as a purified, very bright-white mass. In their striving for a higher consciousness, the alchemists practiced "dying and becoming," as they familiarized themselves with certain processes which they contemplated. In that way, they continued the traditions of the old mysteries, experiencing a death and resurrection.[18]

Lead and Silver as Representatives of Certain World Principles

In our study of mercury we saw how one of the main metals has properties which allow it to be considered as a representative and carrier of principles active in the constitution of the universe. If, from the same perspective, we now compare silver with mercury, with which it has some resemblance, its ability to form surfaces immediately stands out. This silver surface brings about a nearly ideal separation between two regions, and yet of itself it is really nothing. The metal stretches to the extreme boundaries, shows its surface and the surrounding.

The repeated appearance of distinct boundaries is characteristic for nature in its present state. Let us imagine a stone in water: within the boundary we find the crystalline mass of stone, and just outside we find the mobile water. Or, let us think of the surface of water. We can hardly imagine something more delicate and less separating, and yet it separates two completely different worlds. And just at such boundaries we often find the phenomena of reflection. Who has not experienced the fairylike view of a quiet stream in the woods? If we view the surface of water from within the water, we discover a mirror surface which is clearer than the best mirrors made of silver. A quiet water surface, seen from below, gives the impression of an extended, wondrous silver sheath, with a round, clear opening in the middle which has a colored edge.

Boundaries also play an enormous role in the life of organisms. We can even say that in large part life involves surface phenomenon. Even the most delicate organism floating in water will always show clear contours, by which it separates itself from the surroundings. Here we also find a completely different world within the boundary from that without. That is the case in a deeper sense than that of the stone and water, or of water and air. Although the substances within the body are only partially different from those without, there is a fundamental difference: within the skin, the processes proceed in such a way that they serve the organism; outside the boundary, they obey laws which hold for the inorganic substances within the terrestrial whole.

We can now see that the ability of silver to form an amazing, magical surface points to the fact that silver is a very one-sided carrier of this tendency in nature to form boundaries. It is related to conditions which we find in living beings when they are young. Every germinating and growing being not only permeates every part of its body with its life forces, but it continuously pushes the outer world further back. It thus becomes comprehensible that, a piece of silver, more than any other metal, through its plastic coherence and its great conductivity forms a coherent whole, whereby every part quickly takes on the state. More than is the case with an old organism, in the young organism the emphasis is especially on preservation of the internal cohesion of the parts to each other and to the whole, albeit on a higher level than in the case of silver.

If, once again, we compare the properties of lead with those of silver, certain relationships will become clearly evident. We have seen that a piece of silver will form a cohesive whole, more than is the case for other metals. In contrast, a piece of lead is much less of a coherent whole than other metals. If we tear a piece of lead, it offers little resistance and is easily torn apart. Furthermore, lead has only a relatively low heat conductance, so that it is possible for a piece of lead to be quite hot at one end and cold at the other.

Lead is less able to withstand outside influences than silver. On the contrary, its surface is open to the influence of many substances. However, it then reacts by forming a solid layer, which then covers and protects the underlying mass of lead. It can be easily shown that these properties point to a kinship with those forces operating in the living organism, which are the cause of aging. How does aging manifest itself in an organism? In particular we see that the coherence of the parts in an aging organism gradually tends to loosen. In many cases dead layers are formed. Such masses tend to remain in the service of the organism, e.g. as protective layer, although seen from the perspective of the active forces, they are more or less subjected to those of the environment. That means, however that the influences of the environment gradually become effective within the boundaries of the organism. Clear examples of such masses are the bark and wood of trees, the horns and antlers of the ruminants, the shells of the mussels.

We can also demonstrate the kinship of lead with the forces that bring aging in the following way: as we have shown previously, lead as the special ability to retain imprints of events which have happened in the past. We find something similar in old organisms. The dead masses secreted by the plants, can often be used like a history book. The annual rings of trees not only tell us something about number of passages of seasons during its development, but also much about the climate. From the form of a grown plant, we can often read its whole development, because what developed in time remains present in space.

Important relations become apparent when we study the development of the crust of the earth. Regarding lead against this background can even lead us to a significant insight into its essence. The earth has emerged from mobile primordial states which were thoroughly heated by warmth and irradiated by light. Only slowly did the solid substances become as dense and rigid, as dull and abandoned by light and warmth, as we know them. In the past, light and warmth acted as all-dominating powers. Then, so to speak, they discarded the substance and let it become dominated by their enemies, gravity and inertia. It is also important to note that especially in that part of nature, where the original glowing mobility came to the greatest rigidity and dullness, in the crust of the earth, we find that the past has been retained most strongly. In all rocks and minerals we find traces of past processes.

Every time we condense a substance by cooling, which has become a glowing liquid or even a vapor, something like the formation of the earth's crust is repeated. It is possible to use lead to make a nearly exaggerated and one-sided picture of this rigidification and dulling at the transition from the initial phase to the preliminary end phase of earth evolution. It is easier to liquefy and even vaporize lead by heating than is the case with most other metals. On cooling, it will immediately cover itself with a solid layer, become dull, and it will combine more strongly with gravity and inertia than most other metals. And more than other metals, it also has the property of storing what has once occurred.

In every respect, lead is really orientated towards the past, towards that which has occurred previously. For example, it is remarkably and strongly determined by the principle of inertia. Inertia in a mechanical sense is something of extraordinary interest. Every movement, in so far as it occurs under the influence of inertia, points to an event which preceded it. Inertia thus means geared to the past. An inert body will always try to maintain or retain what it once received as impulse. In this sense, we might well say that the chemical behavior of lead is an expression of inertia. This behavior also indicates the tendency to hold onto the past.

In summary, we can say: at various levels in nature there are forces which tend to direct evolution to an end phase, which retains what once was flowing movement as a rigid image. This applies just as well to living organisms, especially plants, as to the earth as a whole. Lead is related to these forces.

In its most prominent expressions, lead is completely and one-sidedly determined by a principle which only manifests itself clearly in living organisms and the earth, during the later stages of development. Summarizing we can say: the properties of silver relate to those of lead as the properties of the young organism to those of the old.

The Importance of Lead and Silver for Civilization

To bring our discussion of silver and lead to a conclusion we must mention that both play a decisive and characteristic role in our society. We need them for the most important tools and technologies, which underlie the expansion of our culture. Lead has been most important for the printing of books. In photography, silver has always played a significant role.

Printing has made it possible for the thoughts of a person to be transmitted directly. Not only can we pass our thoughts on to people in distant places, but also to those living many years later. However, before that is possible, we must "catch" the living flow of thoughts into words. These must be written down as letters and then printed and copied. A living stream is thus brought to stillness. If we really consider everything that goes into producing and making a book available for everyone, we will see that we are dealing here with a process of abstraction and drying up. Thus, reading the frozen word, in the right manner, is an enlivening process. It is remarkable that in a picture which is thought to be one of the oldest depictions of a printer's shop, we see some skeletons. They help the clients and work the presses. Apparently the makers of the pictures knew or had an inkling that through printing, death was being introduced into our culture.

The alchemists called lead "Godfather Death"; they depicted him as a skeleton with scythe. That is indeed a very true image, for in both the poisonous and the healing action of lead, we can always recognize these powers of aging and death. However we view lead, be it from the chemical and mechanical point of view, or from the cultural one, we will always find that lead is involved in a process of bringing to an end or rigidifying a movement, of holding fixedly to something from the past.

Silver presents us with the opportunity to produce images of objects or occurrences quickly and in large numbers and present them to people who otherwise are not in a position to view them. The reason they can't see them could be their spatial or temporal distance; however, it is also possible that ordinary eyes are inadequate. For example, silver has been used in reflecting telescopes to produce images of stars and galaxies which are hidden to the unaided eye. Usually, photographs are taken of these images, again fixed with the aid of silver. These photographs are often used scientifically, printed in books or journals.

Of course, photographs provide a wealth of facts and a real abundance of impressions of earthly objects, which, because of distance, would otherwise never be available to this degree. We can make the same observations about film, in which images are awakened to a semblance of life. Silver is a kind of "magic window" in which we not only see the our immediate surroundings but the whole world. However, we are continuously in danger of misusing photography and especially film, as it is very easy to create a world of hallucinations and illusions, which represent nothing real and only serve to draw us into playing with our own personal emotions and desires.

Gold

Gold in the Metallic State

Gold distinguishes itself from the other metals we want to discuss here just by the fact that it is hardly ever found as an ore, but above all as a pure metal. However, it is found in association with other metal ores. Thus, metallic gold is already completely at home on earth, as it is presently constituted. The density of gold (density = 19.3 g/cm^3) is far greater than that of the other seven metals; it is one of the most dense metals there is. Only platinum and some platinum metals have a greater density (osmium 22.4 g/cm^3). It is even 1.7 times heavier than lead.

Gold surpasses all other substances in its ability to be rolled and hammered into a gold leaf with a thickness of only 0.00014 mm. It is also possible to draw out a gold thread until its diameter is only 1/500th mm. We cannot see such a thread with our unaided eye; we can only see it as a sparkle of light in sunlight under the right conditions. A 2 km long gold thread weighs only a bit more than 0.1 g.[19] From that, we can calculate that a 1 cm^3 chunk of gold will produce a thread 388 km long (using a diameter of 0.002 mm)!

The surface of gold is easily burnished and shows a strong reflectivity (light reflection 85%). With the exception of silver, this is the highest reflective index of all other substances in nature. Gold can not be attacked by any substance which is found in nature, not even by sulfur. Its ability to conduct heat and electricity greatly exceeds that of other substances, except silver and copper.

We have come to regard silver as a near perfect metal. In gold we find that some properties are more enhanced. However, at the same time a contradiction with respect to the essence of metallic nature appears, and that is its color. Pure gold has a beautiful, warm-yellow sheen. Therefore, it does not reflect the surroundings with the same purity as does silver, but imparts to all its images a sun-like cast. On the one hand, its surface is even more impenetrable than that of silver; on the other, it reveals something of itself through its surface. It answers light with a color, which itself is very strongly related to sunlight. More than any other substance, this one has combined with the forces of the earth, yet responds to illumination by sunlight with a sunny shine. In contrast to silver, gold will strongly glow on heating. Liquid gold emits a delicate green color.

What does it mean that we can hammer pieces of gold to a thinness of 0.00014 mm? It indicates that this substance, although never losing its connection to light, can be brought to a state of near weightlessness through pure mechanical processing and thus enhance its relation with light. To achieve this we start by rolling a piece of gold into a thin sheet. Then this sheet is cut into small squares of 1 cm^2 each. These are then placed in square containers with pieces of paper or vellum between each square. They are then hammered carefully until every one of them is extended to a square. Every sheet is again cut into small squares and again hammered. This process is repeated as often as necessary. For the last stage very soft animal skins are used. In this way, it is possible to extend a piece of gold indefinitely, until it is hardly more than two huge shining surfaces with nothing between. It is no small matter when we consider that 1 cm^3 of gold can be hammered into a sheet of 100,000 cm^3 or nearly 10 m^2. If we lift a piece of gold leaf up, then we can let it rest on air as it were, so slowly does it fall to the ground. By holding a gas flame under it, it shoots upwards again. If it is lying on the table, it will flutter up at the smallest breath of air.

These facts indicate what is important for gold. It takes on the "metal cloak," which generally turns away from light, however never losing contact with it. It penetrates deeper into gravity than most other substances, and yet it is easier to make it weightless and lift it again to light than for any other substance. It is characteristic for light that it makes the most of indefinite expansion. Now if gold surpasses every other substance in its ability to be expanded, then here also it demonstrates an affinity for light surpassing all other substances.

In its being, gold has both a deep connection to the realm of light and to that of gravity; it is so all-encompassing that it can embrace the largest polarity in nature. It does not need to choose between light and gravity as most other substances do.

We have hereby also characterized the mechanical behavior of gold. Gold is capable, both with respect to its essence and its relation to space, to connect that which is furthest apart and to span the widest and deepest crevices. It is a fact that gold can maintain its inner coherence under circumstances when most other metals would have been ripped to pieces. The cultivation of inner "coherence" belongs generally to the metal state and gold has achieved the most in this respect.

Comparison of Silver to Gold

By discussing the properties of gold in this way a further light is shed on the essence of silver. It is as if silver is on the right path to accompany gold, but remains stuck half way there. For example, although silver belong to the heavy metals, its specific density is only a little more than half that of gold. It can be hammered into extremely thin layers, yet the thinnest silver foil is still nearly 20 times thicker than the corresponding gold foil. Although its surface is very noble, silver can be attacked chemically by a few natural substances. And as the color shows: silver is very shiny and thus remains at the boundary of metallicity.

To a large extent the same comments apply to the ductility of silver as were given for gold. Here also an extraordinary extension occurs. For the most part gravity is also conquered here, so that silver is also lifted into the realm of light. In the process, the ability to reflect and form images only increases. Both before and after this process, silver continues to reject light; it has associated itself with the great polarity of light and gravity, but keeps them strictly apart.

Gold reconciles what silver keeps separate.

Gold Chemistry

We have established that there is no one substance in nature which chemically attacks gold. If we want to dissolve gold, we can best use "aqua regia," a mixture of hydrochloric acid and nitric acid. We can easily make the action of aqua regia clearly visible by placing a piece of gold foil on the liquid surface. If dissolving proceeds too slowly, slight heating will help. We can then observe how the gold foil becomes ever thinner and more transparent, until in the end it is completely used up. The liquid will have taken on a yellowish color. Basically, we are dealing with the action of chlorine, which is released through the action of nitric acid on hydrochloric acid. If we hold a piece of gold foil in an atmosphere rich in chlorine gas, it will also be quickly attacked, especially if it lies on a moist piece of filter paper. This is also easily demonstrated.

It is a matter of course that gold can easily be reduced from its compounds. By using a solution of gold(III) chloride and carefully following the instructions of the lab textbooks, it is possible to reduce the gold and produce a deep violet, wine-colored solution. Even with a dilute gold (III) chloride solution, we get a strong color—we have here a colloidal gold solution. Under the ultra-microscope we can distinguish innumerable gold particles as small dancing tinsels against a black background. By using special techniques it is possible for gold to be taken up by glass in colloidal form. One part of gold in 100,000 parts of glass will still produce a strong, bright pink color.

Again, gold demonstrates its ability to disperse itself very finely, again in a marvelous interaction with light.[20] Chlorine belongs to the halogens; other halogens will also chemically attack gold. We have learned to know the halogens as those substances which oppose oxygen and thus oppose the chemistry orientated to the sun.[21] Gold as metal of the sun is apparently especially liable to chemical attack from those substances which oppose the sun.

The Principle Represented by Gold

We have learned to know gold as a substance which, more than most others, is bound to earth, and at the same time is predisposed to be exceptionally open to light .

In the case of silver the essence of the metallic state becomes distinctly apparent insofar as the separation of light and matter at its surface is nearly complete.[22] In its ability to reflect light, gold is only slightly behind silver, and yet it responds from within to the illumination by sunlight. It resists the complete separation of light and matter, by reflecting a color having the strongest kinship with light when illuminated by it.

If gold is strongly heated, it starts to radiate with a green color. It, thus, can radiate light, something which silver is hardly capable of, and other metals can do only to a slight degree. Only copper surpasses gold in its ability to radiate light and colors.

By using special processes it is possible to make very thin and extended sheets or foils of gold, more so than with any other metal. This means nothing less than that it is easier for gold than other metals to be transformed into a state in which light-forces would bring any substance, if they could act unhindered. Light always brings about a refinement and extension of matter.[23] The most delicate and delightful examples are to be found for example in the magnesium silicates.[24]

Where in nature at large do we find the place where the most intensive interaction between light and matter takes place? On the surface of the earth. Without the sun, the surface of the earth would be absolutely rigid, bare and colorless, as indeed is the case during the night at the North Pole. Through sunlight colors starts to shine, and the earth answers with various cyclic processes and with the start of life. Especially through the plants, substance is offered to the sun and woven with the aid of light into something new. In this encounter, it is as if the earth wants to imitate the sun as much as possible. Even matter starts to shine a bit.

Gold is the substance which has the predisposition, to a very large degree, to imitate the action of the sun and shine forth.

The First Use of Iron and Copper as Tool-Metal

In this discussion of iron and copper we enter into the area of tool engineering proper. Thus, we will begin our discussion with a brief history of the extraction and usage of these metals, as well as with a sketch of the development of tools. The history of tools is also the story of the use of fire. In the Stone Age, the usage of fire for the preparation of material was hardly practiced. One was confined to using wood, bone, stone, and such materials. Gold was the only metal that could be used, since the use of fire was not necessary for its extraction and treatment. We could, from an engineering perspective, also call the Stone Age the "golden age" or the "age of gold." Gold was hardly used for making implements, but mostly for making jewelry and objects used in religious ceremonies.

The first metal used for tools and weapons was copper. Copper can be extracted without the use of a lot of heat. Copper prepared in this way could be forged in the same way as gold.

Later, bronze was produced; it is an alloy of tin and copper. Tin can also be manufactured relatively easily by melting from its ore. It is also easily alloyed with copper. Bronze objects can be cast, which is not possible with copper. Moreover, it was possible to forge old-style bronze alloy. In any case, fire plays a more important role in the manufacture and treatment of bronze than in the case of copper.

At a later date, it became known how to extract iron from its ores. To this end it was necessary to use complicated ovens working at a higher temperature, using charcoal as fuel. Also, for the treatment of iron it was necessary to use fire, as it is easier to forge hot iron than cold. In later times, the giant blast furnaces that we know today were slowly developed.

Iron as a Metal: The true relation between the human organism and technology

Pure iron shows a nice luster if it has been burnished properly. Its ability to reflect light is somewhat less than that of lead and, thus, much less than that of silver. Iron is, thus, not strongly related to light. Its conductivity for heat and electricity is much smaller than that of silver, gold, and copper. Its magnetic properties, however, are great, and iron far surpasses all other substances in this property.

Through alloying iron with other substances and through various treatments, it can acquire many different mechanical properties. It can be brought into such a state that it becomes very hard, or else very ductile, very elastic, or resists bending, etc. It is thus possible to make it meet the various demands that determine the material a tool is made from. We can best describe this property, which of course is so very important for engineering, as follows: Iron has the special capability to be treated in such a way that it forcibly maintains its inner coherence or shape while being used as a tool to work on the surroundings. These have to retreat, or they have to acquiesce and be transformed.

Gold also demonstrated an all-surpassing capability to maintain its inner coherence. However, this ability did not depend on warding off; gold yields easily and gives way to external mechanical influences. In the case of iron everything is based on force and defense.

A great difficulty with the use of iron is the ease with which it is attacked by atmospheric influences. The speed and thoroughness of the rusting process for iron surpasses that of all other heavy metals. No sooner has the metal been smelted from the earthy state of its ore, then it plunges again without restraint into the earthy state of rust. A brightly shining, strongly coherent substance changes into a brown colored, crumbly, and encrusted mass.

This phenomenon prompts a comparison of iron and lead. Lead yields to any mechanical influence. However, in the case of rusting and other chemical reactions it covers itself with an impenetrable coat against further influences. Iron is without defense against the influences of the environment as seen in rusting and many other chemical reactions; however, it can powerfully resist mechanical influences.

In many respects, iron is closely related to the earth itself. It is found in the crust of the earth in extraordinarily large amounts.[25] If we magnetize a piece of iron, in a sense it becomes an image of the earth, in that it develops two magnetic poles. During rusting it returns as fast as possible to the earth. And just this metal, above all others, can become the carrier of those properties which we need in our tools, so that we can process the substances of the earth and can engage and control the forces of the earth.

We will now briefly look at the diversity of iron technology with the aim of finding some guidelines to understand technology as a historic phenomenon. Many misconceptions abound about the essence of technology and, in particular, about the relation between the human body and technology. Humanity is so impressed by its own achievements in the area of tools and machines, that it forgets more and more the all-surpassing wisdom of the human body. There are many people who think of the human body as nothing more than a complex machine. Only when we had found how to construct strong but light bridges, did we discover that in those parts of our skeleton (head of the femur) which are subject to large stresses, our body uses exactly the same principles. Thus, we came to understand our body better with the aid of experiences and insights gained in engineering and building. Something similar is always the case. Careful study brings to light that in technology, the principles in our body are used on a larger scale, but that our body uses them with much greater perfection. The only correct conclusion would be: We can only apply part of the principles which have been realized comprehensively and perfectly in our bodies since time immemorial.

Already in the last century this conclusion had been reached and a series of consequences correctly drawn from them. Dr. Ernst Kapp published a book in 1877 with the title: ***Foundations of a Philosophy of Technology***.[26] In it we find the sentence: "Thus the mechanism is the torch to illuminate the organism." Also of eminent importance is the concept of "organ projection" which he introduced. All our tools, with respect to their construction or function, are copies of parts of our bodies. Whoever constructs or invents a tool will generally be unaware of this relation or know very little of it. Thus, at the foundation of engineering there lies an unconscious, outward projection of parts of the body or bodily processes. The tools are mostly strongly simplified in comparison to the bodily original, and specialized at the same time, so that for specific objectives, much more may be achieved than with any part of the body.

Examples of tools which relate to the hand, and thus serve to enhance any activity of the hand in a one-sided manner, are: fork, spade, rake, chisel, hammer, etc. Later, such tools often developed into machines. They then function more or less independently of the hand, while previously they had the task to directly support the hand in its activity. In the case of the "organ projections" which relate to the hand, we are often dealing with tools which help reshape the environment. These are "interfering" activities. Machine parts can also be images of arm and hand, i.e., the drive-shaft of a steam engine.

Devices which strengthen and increase the function of the legs play a very large role in our lives. And these especially tend to grow to a gigantic scale. For example, we have a large succession of devices and machines which enhance walking: bicycles, motorbikes, cars, sledges, ships, etc. Then we have those appliances which enhance and support standing: bridges, viewing towers, etc. Countless iron constructions i.e., street-lamps, radio masts, the supporting parts of a railway station, could be considered as standing devices. In a building such as the Eiffel tower, the ability to stand has been projected into the outer world and blown into gigantic proportions. The examples mentioned are only a small selection. Every part of the human body is copied, e.g. the mobile connection between head and neck. At some stage the systematic study of the world of engineering as a spread out or even dismembered human body is a topic which should be taken up.

Since some time, iron has been used for functions which resemble those in our skeleton. Before iron was used, wood was utilized. In the case of this kind of "organ projection," quite a different material was used from that of the body. Nowadays that has changed. Instead of iron reinforced concrete is being used more and more. We have thus introduced a chalk-like substance. In this case the principle of "organ projection" has been realized in such a consequent way that not only follows the construction principles, but it even copies the body's substances.

To conclude, iron offers us the possibility of spanning large distances or even to bridge them to a greater extent than other metals. In addition to transport, we can call to mind iron constructions like bridges or roofing, etc. We have here a property of iron which, in nature, is only clearly manifested in the trajectories of meteors, yet is one of the most essential of iron. As is often the case, such a property only is brought to the fore through engineering. The tendency to measure itself with the great, something world-conquering, belongs to the essence of iron. Whoever lets him/herself be inspired by iron, will never admit that something one wants is impossible, because it is something too large. Iron has something all-embracing, a desire to combine with everything.

If we compare the movement of the planet Mars and the orbit it describes across the heavens, with the movements and orbits of the other planets, then we will discover that we are dealing here with a clear projection of heavenly relations.[27]

Copper as a Metal - Copper and Iron

Copper is often found in nature in combination with iron. In technology it is even used in conjunction with iron. Iron and copper complement each other in many ways.

The most remarkable property of copper is its unusually strong color. We have seen that through its color, gold actually penetrates the boundaries of the metallic state. The ability to reflect strongly and thus to ward off light, and the property to radiate yellow light on illumination, are in opposition with each other. Notwithstanding gold's extraordinary specific density and, as a metal, that it will reflect light, it has a kinship with light. The surface of copper is much less noble than that of gold and its reflectivity is significantly smaller (73%). However, as far as the intensity of the color goes, copper far exceeds gold. Under certain conditions of illumination, copper will respond with a firey yellow-red glow. Here we do not find a modest harmony as a result of an admirable balance, but an intense thrusting forward of oneself. For the onlooker, it is lucky that copper cupolas and roofs quickly loose their brightness and turn green, because the blazing copper color against the background of the blue sky, produces such an overpowering beauty with which the eye can hardly cope. The color transition from the reddish metal to the green patina is a mighty one.

If we compare copper with silver, it becomes clear that copper has very little tendency to hide behind mirror images. On the contrary, the images are not very distinct because they are nearly overpowered by the color of copper.

Also, on being heated copper behaves quite without reservation. When the temperature has risen high enough it starts to glow with a green color, which is more intense than the case of gold. A bewitching image is created when the metal is molten, and the green liquid is interspersed with yellow glowing platelets of oxides, which are swirling around.

Copper is often used in a relatively pure form. It is very ductile, much less so than gold or silver, and, thus, it can be hammered cold. The coppersmith can hammer various objects from a piece of copper. In this way we can produce large kettles but also small utensils, such as vases with narrow necks, or even jewelry.

As regards conductivity for electricity and heat, copper is close to silver. A very large amount of copper is used for electrical leads. As a result, a wide network of copper is produced. We have here an uninterrupted, coherent system of conducting material. External electrical potentials are continuously being generated, and just as ceaselessly, copper cancels these, being a conductor.

The ductility and conductivity for electricity and heat are properties which, in the case of metals, are based in general on the tendency to form a comprehensive whole with an inner cohesiveness. In the case of gold, silver, and copper, this tendency reaches a peak, but in different ways.

At this stage we could discuss further aspects of copper alloys. However, we must restrict ourselves to a few facts, as that suffices for the following. There is a whole range of metals with which copper may be alloyed, with tin forming bronze, zinc forming brass, with gold, silver, with aluminum, etc. However, it is hardly ever alloyed with iron.

On the contrary, we very often find a different kind of cooperation between iron and copper in technology. They are used as materials for different parts of one device, or for different devices which belong together. Iron machines are often completed by small copper parts. The strong tracks of the railway are often accompanied by the copper wiring of the telegraph or by the copper wiring of the electrical leads. Because of its high tensile strength, copper is more suitable than any other metal for making high pressure joints, such as in autoclaves, or as guide rings in canon. Copper, thus, is of great importance to machine and weapon engineering.

In this context, the construction and functioning of electromagnets are interesting. In principle this consists of insulated copper wire wound around an iron core. As soon as an electric current is delivered through the copper, the iron becomes magnetized and exerts a strong force on its surroundings. In other words, as soon as we allow copper to unfold its equalizing, relaxing activity, iron is brought into a state of tension. If iron is magnetized externally, then an electric potential is generated in the copper wire, which can result in an electric current.

Why is the electromagnet one of our most useful and powerful power sources? In this device the relation between the natures of iron and copper has been applied and utilized in the most consistent manner. The iron core and the copper wire are perpendicular with respect to each other at every point, and yet they retain even purely spatially the most intimate relation. The ability to equalize potential differences and the ability to generate power, are shown here to be opposites, which, however, always appear together.

Here we have one of the most magnificent examples of a rule, we have often formulated: Where technology reaches a culminating point, certain laws of nature are clearly demonstrated. Only if we work in total harmony with the most fundamental laws of nature will we be able to achieve anything in engineering. Often we are dealing with laws and principles of nature which remain completely unknown to the engineer and which we hardly ever find articulated.

Iron in the human body and in Nature—
Chemical properties of Iron

Our whole body is permeated by iron. It forms the basis for the red color of hemoglobin. It also plays an important role in the production of pigment in our skin and hair. Thus, it participates in two areas in color formation. Despite the fact that the amount of iron in our body is not very large,[28] it contributes substantially to our appearance.

The iron in blood is involved in strong movement. It incessantly moves through the circulatory system and reaches the most distant parts. At one stage it moves with the blood into the lungs, to come into contact with the air via the vast surface area of the lungs, then again it contracts when flowing to the heart. Now it flows with a bright red color into the seclusion of the body, to return sluggishly and with a deep red color into the heart. It takes oxygen with it on its travels, and thus iron contributes towards life being continually re-awakened and sustained. While oxygen is given off in the farthest parts of the body, carbon dioxide is taken up to be released in the lungs

Iron plays a second role as mediator between oxygen and the finer processes involved in tissue respiration.

We have previously acknowledged that we must see the properties of the substances of the circle of chemical elements[29] in relation to the most important tasks they fulfil in nature. When we know the most important realm in which they occur, then we can immediately deduce their properties. In general, this rule is not of great importance for the heavy metals. However, it does apply in the case of iron. In the same way that magnesium appears in the center of chlorophyll, hemoglobin in human beings and animals relies on iron. Just as we have to consider the properties of magnesium in relation to the role of chlorophyll in the plants, so, in the case of iron, we should regard the human body and especially the blood, as that part of nature from which to deduce the properties of iron. In the way iron occurs in nature, and through its chemical properties, we are constantly reminded of its role in human and animal respiration. Iron compounds are about as universally distributed in nature as blood is in the human body. It is found everywhere in the crust of the earth, in water, and even in the atmosphere as meteor dust. The meteors, which penetrate the atmosphere from beyond, insofar as they contain iron react with oxygen and descend to a large extent as a rain of rust (e.g. iron oxides).[30]

A very extraordinary circulation of iron, which to a large degree resembles the processes in the human body, is found in water and especially in oceans and seas. Iron sinks to the bottom as insoluble iron (III) hydroxide [$Fe(OH)_3$], which is oxygen-rich, and poor in carbon dioxide, and rises as iron(II) bicarbonate [$Fe(HCO_3)_2$], which is relatively poor in oxygen and rich in carbon dioxide.[31]

If we leave a piece of iron exposed to the elements, it immediately reacts with oxygen, assisted by water and carbon dioxide.[32] This process is not hindered by the formation of the layer of rust, but rather is promoted. It goes on steadily until every particle of metal, however deep it may lay under the earth, becomes brown and earth-like. The fact that oxygen penetrates also here into the deepest layers is the more remarkable, as the other seven planetary metals without exception resist rusting in some way. Copper, which resembles iron closest in the process of rusting,[33] protects itself through the impenetrability of the layer of rust once this is formed. Even this taking in of oxygen in the density of matter can be seen in relation to the task of iron in our blood.

Iron rust is predominantly iron hydroxide.

By heating iron, an oxide is formed, hammer scale Fe_3O_4. Initially, we see various nuances of color shimmering across the metal surface, which quickly becomes dull and very dark. These colors are called oxidization tint. The easiest way to demonstrate these colors is by putting a razor blade in the flame of a burner. The region being heated quickly goes dark, while on the boundary between the dark and bright region, a band emerges which has the color sequence of Goethe's inverted spectrum.[34] A very surprising phenomenon! In a chemical reaction, colors come about which obey the laws of spectroscopy.

Everyone knows that when sharpening knives or when welding two pieces of iron, we get a shower of sparks. Iron surpasses other metals in its ability to produce sparks. When it shows itself in the form of a meteor, then it generates a sparkle light phenomenon of cosmic proportions.

In experiments we can enhance this property to amazing phenomena. We can achieve this by blowing or sprinkling some iron powder into a blue flame of a Bunsen burner in a dark room. We immediately see a graceful play of short lived flashing lines of light. If we use a somewhat coarser powder, then the lines will even branch out in wonderful ways. In the case of such experiments we should take our time and vary the way of executing the experiment. We will then see how everything becomes more beautiful.

Of the seven metals, iron is the only one which will dissolve in various diluted acids. In the case of the others, we encounter varying oppositions: they may be too noble, or the resulting salts are insoluble, etc. Also, in this respect iron shows the least restraint. Adding iron powder to sulfuric acid will make the liquid bubble powerfully, while hydrogen gas is being produced. A green solution of iron (II) sulfate (ferrous sulfate, $FeSO_4$) is formed. (When iron takes on the valence 2, we talk about iron (II) compounds; an older designation is ferrous compounds.)

If we dissolve iron in dilute nitric acid, which has strong oxidizing properties, then we get a red brown colored solution of ferric or iron(III) nitrate $[Fe(NO_3)]_3$. (When iron occurs in the state of valence 3, we talk about ferric or iron (III) compounds.) Most metals can occur in two or even more valence states. In the case of iron, this is of exceptional importance, as both in nature and in chemistry, iron continuously moves between these two valence states. It might be helpful to express this fact again but in a less abstract way. If we look at the mass ratio between iron and the substances with which it reacts, then we get the following result: in the case of the iron (II) compounds more iron reacts than in the case of iron (III) compounds with the same amount of substance.[35] This oscillating of iron is apparently related to its task in respiration. We can no doubt state that at the surface of the earth iron is found in the valence state 3, as a reddish compound.

We can demonstrate the oscillation easily and beautifully with the aid of the color change. First we make a solution of an iron (II) salt. For example, we can dissolve some crystalline iron (II) sulfate or even better iron (II) ammonium sulfate (Mohr's salt, ferrous ammonium sulfate) $(NH_4)_2Fe(SO_4)_2 \cdot 6H_2O$ in some boiled water, which has been made slightly acidic. We can get better results if we dissolve some iron powder in dilute sulfuric acid, because in the first case it might be possible, through the interaction with oxygen, to find minute quantities of iron (III) sulfate. We then add some boiled and subsequently cooled sodium hydroxide. If everything has been carefully prepared, we get a white precipitate of iron (II) hydroxide. However, mostly we will get a greenish or even a black precipitate, because the smallest amounts of oxygen cause a color change. If we filter the precipitate, we will find a green jelly-like substance on the paper. By unfolding the filter and leaving it open to the air, then the color will slowly change to a bright yellow orange. Iron (II) hydroxide is being oxidized to iron (III) hydroxide. Something even more beautiful may be achieved when we produce the precipitate in a beaker glass and leave it quietly during many days. We will then see how the brown coloring slowly penetrates into the depth of the solution. Impressions are formed, which can remind one of a beautiful sunset.

As students are mostly too impatient to follow such slow processes, it is possible to work towards getting results more quickly, but with less beautiful results. In a test tube some iron (II) hydroxide is formed and then shaken powerfully. It is possible to change dissolved iron (II) salt quite quickly into iron (III) salt by adding some nitric acid. By using a medium beaker or Erlenmeyer flask, it is possible to achieve impressive experimental results. Within the light green solution, a dark coloring, like a threatening thunder cloud, forms almost immediately with the addition of some nitric acid. Shortly afterwards, the solution will effervesce and brown nitrous vapors will be released. The cloud will quickly disappear, and the color will become a transparent brown-red.

The opposite path is pursued if we add some fine iron powder and sulfuric acid to a brown solution of iron (III) sulfate. After the solution has effervesced strongly and has settled again, the solution is filtered. The resulting solution is colored green due to the formation of iron (II) sulfate. The resulting (nascent) hydrogen has accomplished the reduction of the iron (III) to iron (II) sulfate.[36]

If we take into account the manner in which iron assists in the uptake of oxygen in our body and how it brings it into the furthest reaches of the body, then we will understand that we are not only dealing here with an energy-giving and life-sustaining activity, but also with a healing one. We can actually exert strong healing actions with iron. How do we discover such actions in chemistry? In the first place it may be remarked that in contrast to other heavy metals, there are no poisonous actions associated with iron. Furthermore, it has the remarkable ability to neutralize the action of many powerful poisons. In gas factories, coal gas is detoxified by passing it through ironstone (ferrungious earth). As a result, we get "Prussian blue" and iron sulfide. In Prussian blue the very poisonous prussic acid is being held by iron and, thus is, unharmful. Iron is also capable of controlling the problematical side of sulfur, which is especially noticeable in hydrogen sulfide. There are processes in which iron and sulfur pull each other down, so to speak. However there are others where they ennoble each other. For example, we can create a frightening image by adding a solution of sodium sulfide to a solution of an iron (II) salt. By doing this experiment on a larger scale and against a white background, the precipitate will spread as a threatening cloud. Pyrite (natural iron disulfide FeS_2), is as it were on the way to developing qualities of gold.

Copper and Iron

Iron

If copper is left exposed to the atmosphere, it will become covered with a dull green patina of basic copper hydroxide.[37] In this respect, it behaves similarly to iron in forming a strongly-colored rust—black copper (II) oxide and red copper (I) oxide. However, the color of copper rust is more or less opposite to that of iron rust. It is of great importance for using copper that its rust is impermeable and thus forms a protective layer. While we have to protect iron constructions by painting them, copper forms its own protection, as it were.

By heating the bright red copper, it turns black because it forms a layer of oxide. By placing this layer within the gas flame, the red color of the metal quickly re-emerges. The oxide is reduced to copper. We can enhance this phenomenon in a large scale experiment. Place a piece of copper (about 10 cm by 10 cm) in such a manner that the audience can see the surface. We then heat the piece of metal from behind, until it has turned completely black. Then, we touch the front of the piece of copper with the flame and wait until the red color has returned within the flame. If we now quickly remove the flame, the metal will again turn black, after having exhibited a quick succession of beautiful colors. If we move the flame slowly across the surface, then it is as if we are pulling the metal color with us. A boundary of strong colors surrounds the red and follows it around. The longer we continue with the experiment and the more often we move the flame around, the more magic-like the play of colors becomes. This experiment corresponds to the one where a razor blade was heated in a flame. Here the colors are in every respect much stronger and more mobile. As with iron, we are dealing here with the inverted spectrum.

Copper dissolves only in nitric acid, not in any of the other acids (except hot concentrated sulfuric acid). For example, we can put some copper turnings into a medium retort (or round bottom flask). On the addition of some moderately diluted nitric acid, the space becomes filled with wildly agitated brown vapors, while the liquid turns a deep blue color. We thus experience again such a mighty color change from the red metal to the blue liquid.

The formation of the pale blue precipitate of copper (II) hydroxide is strikingly beautiful. This precipitate is formed by adding sodium hydroxide to a copper (II) salt solution. We can also reverse the order of addition, and add a copper (II) salt solution to a sodium hydroxide solution.[38] The color of the precipitate can be seen better in the colorless liquid. We get remarkable results by hanging a crystal of copper (II) sulfate in a solution of sodium hydroxide. If the concentration is correctly chosen, pillars of copper (II) hydroxide will form.

If we heat a beautiful precipitate of copper (II) hydroxide, it turns quite ugly—the black color of copper (II) oxide appears. We can use the oxide freshly formed in this manner, to demonstrate the brightening, darkness-overcoming, effect of acids. With a little hydrochloric or sulfuric acid, it will easily dissolve and change into a blue solution.[39]

Experimenting with copper and iron becomes especially interesting if we take the spectrum as our starting point. With iron compounds it is quite easy to produce the main colors of the spectrum. We start off with an orange red solution of iron (III) chloride ($FeCl_3$). Part of this solution is diluted to achieve a yellow color. To a part of the diluted solution we then add a few drops of acid and some potassium thiocyanate (KSCN) solution to produce a deep red color. We get a green color by preparing a solution of iron (II) hydroxide. A few drops of iron (III) chloride solution added to a solution of iron (II) cyanate produces a deep blue solution (Prussian blue). We can get a violet color by adding some iron chloride solution to a tannic acid solution (the old way of preparing ink). If we place the results in sequence at the end of this series of experiments, we are presented with an image which can richly impress us.

Very impressive experiments are achieved when we work, as it were, symmetrically with copper and iron compounds. Place two beakers (or 3-liter conical flasks) over two light sources.[40] Both containers are filled with tap water. In one of them we pour some concentrated copper (II) sulfate solution. After the solution has diffused throughout the flask, the contents will be a transparent pale blue. However, quite slowly a cloudiness will appear which gives off a gentle blue radiance. By adding a little acid, this cloudiness will disappear again. During the dissolving of the cloudiness we can observe an astonishing metamorphosis, which can be enhanced by bringing the liquid into a slow rotational movement, using a glass rod. We are confronted with a picture which looks remarkably like the dissolving of cirrus clouds.

To the content of the second flask, we add some iron (III) chloride solution. Initially the color will be a pale orange-brown. Here also a cloudiness will form, which gives off a gentle brown radiance. Here again, a little acid will dissolve the cloudiness. The play of forms will be less delightful than in the case of copper (II) sulfate, especially of a less delicate design. We now add some sodium hydroxide to both flasks. In the first flask we will see delicate blue colored flakes of copper (II) hydroxide; in the second, flakes of brown colored iron (III) hydroxide will appear. Both experiments are identical with respect to the processes; but with respect to the colors, they move in opposite directions of the spectrum. The appearance of copper compounds could be called aesthetically delicate. The iron compounds seem more robust in appearance.

By adding some acid, both precipitates dissolve. We then add a few drops of potassium thiocyanate (KSCN) solution to the iron salt, and deep red streams of color appear. We add some ammonia solution (NH$_4$OH) to the copper salt, and a deep ultramarine blue-colored liquid appears. By projecting a spectrum and placing the blue colored solution in the light beam, then the whole yellow-red part of the spectrum will be extinguished. The red solution will extinguish nearly the whole spectrum and only a narrow red band will remain.

We now fill two large Erlenmeyer flasks nearly to the top with tap water and freshly prepare a weakly acidic solution of copper (II) sulfate and iron (III) chloride solutions. We then take a yellow crystal of potassium iron (III) cyanide and carefully dip it into the iron (III) chloride solution. A precipitate is formed, which sinks to the bottom as a mass of dark blue colored flakes. For this experiment we need a bright background (thus, a matte piece of glass or an illuminated white surface). We then rinse the crystal and dip it into the copper (II) sulfate solution. We get a similar, but reddish colored precipitate, which will radiate beautifully when placed over an illuminator.[41] In all these cases colors are produced which are contrasting or at least symmetrical for copper and iron, when using the normal spectrum as reference. That applies even when colors appear which are unexpected, as in the case just discussed.

From these experiments we can deduce that iron and copper are closely attuned to each other. Their opposition, which mirrors the relation between Mars and Venus, strongly manifests itself in the realm of color.

Using the flame of a Bunsen burner we can produce phenomena which point to a very special relation between copper and the realm of colors. By blowing some copper powder into the flame, we will observe a green radiating cloud, and not the sparks as is the case with iron powder.

Something really spectacular manifests itself if we direct a strongly burning gas flame onto a heap of copper(II) chloride lying on a stone. Initially, the flame takes on a green color, but quite quickly a strong radiating blue emerges, surrounded by a dark red border. Orange sparks flash through the whole. If we make a little effort we will discover all colors. There is no other metal of which the flame approaches that of copper in richness of color.

If we do the same experiment with iron (III) chloride, then we will see bright yellow streaks which follow the direction of the flame. The impression of pure force is given. There is really no display of color to speak of.

The Principles Represented by Copper and Iron

Copper contradicts the essence of metals to a stronger degree than gold. We have seen how silver takes a certain side of the metallic state to an extreme, namely the clear separation of inner and outer. Copper tries, as it were, to further enhance the beauty of the metal surface, not only by mirroring the surroundings, but intermixing its own strong color, as much as possible. It appears like a glowing protest against the delicate, chaste, self-effacing silver.

Every surface in nature is a place where a delineation takes place. The pure separation of two realms is mostly the main function of such a boundary. However, it is nearly always the case that, as with living beings, colors and patterns occur through which much of the nature of the being comes to expression. In many cases we find an enhancement of this phenomenon during the development of such a being. The growing plant often shows a developmental progress towards a manifestation of the highest radiance and beauty, which is then followed by a relapse to a stage of growth, which has little to offer the eye. Compare the beautiful flowering plant with the inconspicuous seed, with the trunk of a tree in winter, with a rootstock. We encounter something similar in the case of animals. For example, think of a stickleback, which during part of the year appears quite inconspicuous, but during the mating season appears as if it "goes up in flames." Humans (and especially women) often go through a stage when it is as if the whole outer appearance is permeated by a radiance of beauty. In the case of living beings this beautiful radiance occurs when the healthy development has reached an end or at least a culmination has been reached. Everything which partly stood in the background as possibility initially has now, for the moist part, come to development.

The different ways a landscape can appear may also belong to this phenomenon. It can also light up in the most beautiful splendor and then appear quite ordinary.

What otherwise only appears in certain stages of life or as a temporary stage, in copper we find it coming to the fore in quite a one-sided manner. We could say of copper that more than other metals it has the urge to express itself as beautifully and explicitly as possible in the sphere of the senses.

Comparatively speaking, iron has not got a lot to show when it appears as metal. The impression its surface makes on us is a dim version of silver's. The ability to ward off influences, and thus maintain the separation of inner and outer, is weak. Pure iron does not show its mechanical properties strongly. It is relatively soft and yielding. However, as soon as it is mixed with some carbon or even with other metals, it acquires significant properties. A piece of metal will generally yield and change shape when hit by a hammer. A piece of hardened steel hardly looks any different from a piece of iron; as soon as it is hit strongly with a hammer, we are confronted with a surprise. It answers with a ringing sound, and the hammer rebounds elastically, without any change of shape. A wire made from an other metal, when strongly stretched, will expand considerably and quickly break. Only under the greatest resistance will a steel wire expand slightly, and afterwards recoil elastically, regaining its original length.

Properly processed, iron can achieve great feats of power; however, this happens only when the attack comes from outside. Here also we have properties which mostly do not impress our senses strongly.

All the various processes which take place in the world—whether they take place in earth, in water, in air; in trees resisting the storms or in the skeletons of humans and animals as they move—are accompanied by strong forces. Even though this play of forces is impressive, we normally don't pay much attention to, because it is hardly ever observed. In this play of forces, a strong urge for realization, for coming into existence, is also manifest. However, initially we are not dealing here with an impulse to express the inner nature in outer phenomena, as is the case with copper, but with self-assertion with respect to the mechanical influences from the environment, and with forces which make it possible to grasp the surroundings that make movement and change possible. Of course, every substance participates to a certain degree in such hidden force effects. However, iron is the substance in which the principle of such effects of forces is expressed most strongly and one-sidedly.

Justifiably, iron is brought into relationship with the planet Mars. In principle, it has to do with war and battle. Battle is only possible when two powers or two beings are strongly opposed. And at this time, one of the main characteristics of the present stage of the evolution of the earth is that all beings on earth, who are in reality of a spiritual nature, condense and solidify into objects. From this perspective, living beings also are objects. The forces we discussed previously and which are especially represented by iron, are also those forces which preserve this standing-in-opposition of objects.

It is quite possible to perceive a contradiction in what has been said previously and the discussion above. Previously, we have seen that, in a certain way, iron thrusts itself into the sphere of the senses by contributing a lot to the colorfulness of the soil and the living organisms. Now, however, we are told that iron has got properties which are hidden from the senses. Here we must carefully distinguish two sides of iron. On the one hand, materially and qualitatively, it is quite strongly engaged in the composition of the crust of the earth. It is to a certain degree part of the circle of elements. Precisely to that extent, iron, in the form of compounds, contributes a lot to the appearance of nature. However, in the above discussion, we are dealing with iron in its kinship to the heavy metals. In its metallic state, iron, just as other metals, brings a specific principle to expression in nature in a pure and one-sided way, without in any way materially or qualitatively playing a great role in the consequence of this principle somewhere else.

It is worth reflecting that in most cases in industry, in contrast to nature, the metal is used as a substance when the principle which it represents has to be applied. On occasion, we should try to imagine the play of forces in a bridge or in some tool.

In this way, copper and iron show a profound correspondence, in that both seem to be filled with a powerful urge to expression. Time and again we find associated with these metals almost exaggerated, intense phenomena. We can then use them to achieve results which require the utmost effect. However, this urge is directed towards different realms: in the case of copper towards in the development of beauty, in the case of iron towards that of force. Insofar as it is possible to produce very tough materials from copper, it moves a little into the area of iron. Iron moves into the area of copper, in so far as it is able to produce strong color phenomena.

Mercury and Tin as Metals

We have already discussed mercury as the bearer of the mercurial principle. We strongly emphasized the property of liquidity.[42] We will now have to look somewhat more carefully at the material properties of mercury, especially with respect to silver. Mercury is also not very reactive, colorless, and clear as silver. Even though its liquid surface is as smooth as if it had been burnished, its reflectivity is much less than that of silver (only 73%). A remarkable property of mercury is its tendency to close in on itself—the tendency to form droplets. If we drop some mercury from a certain height, it will form innumerable droplets on impact with the ground. The fact that it is not only very mobile, but also is exceptionally heavy (specific density 13.6 g/cm^3), contributes greatly to this phenomenon. When we are able to bring all the drops together, they will flow together into one large whole.

On a clean metal surface like one of gold, mercury will not form droplets, but will form a plane smooth surface. This is the basis of the old (dangerous) "game" of coating pennies with mercury.

The tendency to disperse and flow together again can only be shown clearly by using certain devices. Otherwise, the tendency of mercury to immediately attract dust and dirt to its surface will complicate matters. We will use a 150 to 250 ml conical flask and add so much mercury that the bottom is ¼ covered. We then cover the mercury with a layer of distilled water. Finally, a few drops of nitric acid are added. Mercury will become more mobile than it already is and its surface will exhibit a beautiful shine. By shaking the flask intensely, we will see the mass being ever more dispersed, until in the end we see a gray surface. If we now stop shaking, the droplets will reform the whole liquid mass by fits and starts. This experiment is especially beautiful if we do it in full sunlight, or if we work in a darkened room with one strong light. Initially, we see one image of the sun, which soon breaks up into a thousand ever smaller images, only to finally form again one image.

Mercury always invites us to play. If we try to press on it, it will get out of the way and flow from underneath the finger. If we pull the finger away, the mercury will also flow back. If we hit it lightly, the surface will react showing wave patterns, which slowly will come to rest. Mercury reacts to everything, even to the most subtle of influences. It always yields, is always slightly dispersed, it can not retain any shape, and yet asserts itself emphatically.

By comparing these properties with those of other metals, we get a clear impression of the nature of mercury.

If we leave some mercury in an open container, it will disappear very slowly, but we can't see were it went. If we slightly increase the temperature, it will disappear more quickly. Most metals are substances which because of their weight are strongly bound to the earth. Many of these confirm this relationship by taking up substances from the atmosphere and transforming them into the earth-like state. Mercury very readily takes the opposite route. Already at normal temperatures, it is able to release itself from its bonds to the earth and disperse into the atmosphere.

We have already heard about the strange properties of metal surfaces. Mercury adds a very new and surprising property. Because it is silver-like, we would expect a sharp distinction between inner and outer. However, this metal easily penetrates through its own surface into the surroundings.

If we heat mercury in a closed space (for example in a retort or test tube), than it will also disappear from its location, but at cooler sites it will reappear as drops or as a mirror surface coating. In this case, it penetrates back into its own surface, but now in the opposite direction.

One cannot say that mercury does not distinguish itself in this respect from other metals, if we heat these to a high enough temperature. However, it is typical for mercury to exhibit such properties at low temperatures.

Most metals can easily be dissolved in mercury. Exceptions are iron and some related metals. By bringing a metal into contact with mercury, it loses its isolation. It is slowly absorbed by mercury and loses its own character. Generally, mercury has a great ability not to be restrained by boundaries and rigid forms. If we throw a golden ring into mercury, it will not only sink, but disappear completely. By carefully evaporating the mercury, we can find beautiful sparkling golden crystals. If we place a piece of silver on mercury, it will float only slightly protruding above the surface and slowly dissolve. This property of mercury to dissolve metals is used in the extraction of gold and silver from mineral ores.

However, if we place an iron ring on mercury, it will float half submerged. This is used for example in lighthouses, where the system of prisms and lenses floats on mercury, achieving mobility with a minimal resistance to movement.

The fact that mercury is both liquid and heavy was of great importance for the mercury barometer. Torricelli (1608 - 1647) was the first to achieve a vacuum above his mercury column. Even nowadays mercury is used in high vacuum pumps.

The use of the pneumatic trough[43] for collecting gasses is greatly enhanced if mercury is used.

Mercury also plays a really important role as liquid in thermometers.

It is obvious: mercury is especially needed in those cases where we are dealing with gasses and heat, thus with both higher elements.

The electrical conductivity of mercury is low compared to that of other metals. However, it has become important because it is very high with respect to that of other liquids. Mercury is used when very mobile, quick, electrical contacts are needed, for example in switching thermometers and pendulum for seconds.

Tin

Just as the surface of mercury resembles that of silver, so the surface of tin is similar to lead. The (optical) reflectivity of a burnished tin surface is certainly relatively large (reflectivity 76%), slightly larger than that of mercury and much larger than that of lead. For a time it may resemble silver. It also retains its brightness longer than lead. However, it does slowly becomes tarnished and gray, especially when it is subjected to atmospheric influences. As in the case of lead, tin offers a fairly strong resistance to penetrating chemical influences, yet, strictly speaking, it is not one of the more noble metals.

Tin is fairly soft, but less so than lead. It is not possible to break a rod of tin, because it can be bent backwards and forwards for quite a while, and can even be expanded and pulled apart like clay. It seems as if tin, like lead, will largely yield to influences; however, if we pay careful attention, we will notice something of a slight protest. When we bend a piece of tin, we will hear a delicate cracking or scratching. Jokingly this is called "the tin cry." We are confronted here with an expression of a noteworthy contradiction: tin is simultaneously ductile, and yet crystalline through and through. If we remove the smooth surface layer by rubbing it with some aqua regia, then tin starts to sparkle. Something extraordinarily beautiful is seen if we carefully treat some tin-plate, iron coated with tin, with aqua regia. The layer of tin is so thin that the crystals can only lie parallel with the surface. After etching away the surface, they give an image as of ice flowers on a window pane, only more delicate: gloriously radiating out fan-like, and yet forming a cohesive pattern of delicately glistering crystal shapes. With present day tin-plate, the phenomenon is not so beautiful as previously, as the layer of tin is too thin. Perhaps one should try to coat a piece of iron with tin oneself, or pour a thin layer of tin.

In any case we can get a clear view of the nature of tin, by etching. If we penetrate to what the unassuming outer layer hides, we encounter a beautifully structured inner world, which appears to be totally attuned to a sparkling game with light. Notwithstanding the fine structure, we will not see any colors.

Tin melts even more readily than lead (melting point 232 °C). Thus, it can be easily poured. In the past many utensils were poured from tin. It was especially important for drinking and eating utensils, because it is hardly attacked by moist or liquid foods, and because it is not poisonous. It is absolutely unusable for pots and pans, as it immediately begins to melt in the fire. Soldering with pure tin is, therefore, also very difficult. Nowadays there are hardly any objects made of pure tin, because it is considered unsuitable for use and especially because it is too expensive.

As we are primarily interested in the properties of the surfaces, it is quite possible to use tin as a thin layer carried on another metal. Furthermore, it has the characteristic, and for us important property, that on heating it will easily flow in a thin layer on other metals, and on cooling will adhere. For example, copper cooking pots and leaden water-pipes are used which both have been coated with tin on the inside. Vast amounts of tin-coated iron are especially used. The greatest part of the tin production is used nowadays in making tin-plate. In the past, suitably cleaned sheet-iron was briefly dipped into molten tin. Now electrolytic means are used to apply the plating. At any rate, most of the tin production is still used for the same reason as in the past, namely to handle and store liquid or moist food.

If we want to gain a picture of the nature of tin, then we only need to look at the engineering usage of tin, even though this is very simple, uniform, and at first sight not very enlightening. Tin is most often used in those instances where we want to protect metals from the influences of liquids. Thus, we will find most tin in thin layers at those places where liquids need to be separated from solids. Tin is thus used for a task which is an exact polarity to that of mercury. We have seen how it is especially capable blurring boundaries.

The use of a can made from tin-plate is, therefore, very practical, because iron gives it the strength to retain its shape, while the thin layer of tin keeps the liquid in its place. If there was no tin, then the boundary between metal and liquid would soon become unclear. Iron would be attacked, and the content of the can would acquire an iron-like taste.

Its great mobility and changeability are characteristic for mercury; for tin, through maintaining boundaries its asset is the ability to protect, a certain order.

Such a deeply fundamental contrast between tin and mercury often expresses itself in a remarkably simple way. While mercury can only be handled when we place it in a container, tin is used specifically where a liquid needs to be contained and kept in one place.

Without doubt we may look at the use of tin alloys in bearings. Between the movable parts and the bearing that carries them a thin layer of oil has to be maintained. Here, also, clear boundaries, and sheet-wise ordering must be upheld.

To conclude, let us look once more at a piece of tin as it is produced through casting where a piece of metal protects and hides behind a tough, smooth and, in a manner of speaking, repelling skin, a finely differentiated inner world. Doubtless, it is now absolutely clear that here also everything points to an order based on boundaries and differentiation. In the first instance, outer and inner world are carefully separated. We then see how under the surface, tin is crystallized. This means that we find numerous, delicate boundaries, which enable the crystals to grow in their precise regular forms. In the case of lead we find that properties typical of matter—heaviness and inertia—dominate, and the individual form has very little meaning. In the case of tin we are dealing with order and form.

Decades ago, tinfoil ("silverpaper") was used to wrap chocolate and the like in (nowadays aluminum foil is used predominantly). In a somewhat humorous way, the nature of tin is again expressed – something very fancy, very promising, was being protected and hidden by a layer of tin.

The chemical properties of tin will not aid us in gaining a much deeper understanding of tin. If we heat it strongly, it will burn with a white flame. The negative fact that the chemistry of tin is not very colorful is perhaps characteristic.

The Chemistry of Mercury

Mercury has a curious relationship to oxygen. There it shows a slight volatility which does concur with its otherwise mobile nature. If we just let it stand, its surface will not be attacked by the atmosphere or even by oxygen in high concentrations. Only ozone (O_3) will oxidize mercury at normal temperatures. However, if we heat it above 300°C in air, it will become covered with a red, powdery layer—mercuric oxide HgO is produced. Mercury boils at 357°C. At 400°C the newly-formed oxide dissociates again into oxygen and mercury vapor.

We can easily bring about this dissociation in a Pyrex test tube. If we heat it with a Bunsen flame, the color becomes really dark—and on cooling, lighter. The mass becomes very mobile; the smallest bump will make it move back and forth, as it is carried by the developing gas and vapors. Above the heated (or hot) zone a ring shaped mercury mirror will develop. A glowing splint of wood placed in the tube will burst into flame, because of the production of oxygen. We must be careful to ensure that the POISONOUS VAPORS do not escape in to the room.

Metallic mercury can easily be dissolved in dilute nitric acid; the oxide is dissolved in dilute sulfuric acid. If we want to prepare a solution of one of the salts, we first have to add a few drops of acid to the water; otherwise, a cloudy solution results due to the formation of a basic salt.

By adding sodium hydroxide to a solution of mercury (II) nitrate ($Hg(NO_2)_2$ we get a red, or sometimes a differently colored precipitate of mercury (II) oxide.

Mercury (II) chloride (or "corrosive sublimate"), a salt with strong disinfectant properties, owes its name "sublimate" to the fact that it volatilizes easily. On heating, it will evaporate without first melting, and then condense at cooler parts. In general, metal chlorides exhibit a certain tendency to volatilize; it should not surprise anybody that mercury (II) chloride shows this more than others.

One of the most beautiful precipitates one can produce will come about when we bring mercury (II) nitrate and potassium iodide together. Mercury (II) iodide (HgI_2) is formed. Fill a large conical flask with pure water—if the chloride content is too large, this will interfere—and place it on the illuminator in a darkened room. Then, add some mercury (II) nitrate solution to which some nitric acid as been added. We should now have a clear solution. If we now add drop-wise some potassium iodide solution, we get a beautiful yellow precipitate, which slowly turns a salmon red and finally settles on the bottom as a bright red precipitate. If, before it reaches that stage, we add more potassium iodide, the precipitate will dissolve again. If we drop a rather large crystal of mercury (II) nitrate into such a solution, so that it remains on the bottom of the flask, and we swirl the liquid in a rotary motion, then we get a wonderful dance of rising and falling eddies of colored veils. These veils are exceptionally delicate and fragile. All this depends on the fact that the processes fluctuate around a sort of boundary. There where there is an excess of mercury (II) nitrate, a precipitate is formed; at other sites where there is an excess of potassium iodide, it dissolves again.

It often happens that we get really surprising images when experimenting at such boundaries.

In this case the rhythmic-mobile nature of mercury is conjured up in front of our eyes.

Mercury and Tin

We have frequently characterized mercury as the representative of a certain principle in nature. When we know that one of the main metals is a carrier of certain principles, then it becomes immediately clear that something similar is the case with the other main metals. We have looked for these principles and indeed have found them. It is not too difficult to verify that these principles are subject to a certain order, in that they appear in polarities. That has been evident for each three pairs of metals; only gold occupied a central position. Thus, we can more easily find the principle which tin represents, because we may expect a polar relation to that which is represented by mercury.

In the case of the principle which is represented by mercury, we are dealing with a breaking of boundaries, a bringing into interaction of what is separated, a mixing, a process of circulation. In the whole of nature, water especially represents this principle. The processes which follow this principle would result in a general "levelling" or "sameness" and loss of form, if there were not an opposite principle at work. Therefore, we are looking at differentiation which rests on clear distinctions. It is very enlightening that the whole of nature is permeated with a strict order resting on various layers. Earth, water, air and even warmth are found in layers one above the other. If we look at the sky, it is clear that even in formation of clouds there is a tendency to form layers. In the crust of the earth, we also find strata built up. Perhaps we should even look from this point of view at the structure of the kingdoms of nature, which in their internal relations exhibit a hierarchical arrangement. Tin must be related to this arrangement, whose wisdom is always impressive. Perhaps the best way to succinctly characterize this principle, is as "order through differentiation and hierarchy." As the use of tin in technology indicates, it keeps everything in its proper place.

It goes without saying that if the only principle at work in nature was that of tin, nature would rigidify. The principle of mercury mobilizes what has become rigid by breaking up the boundaries and introducing lively interactions.

Quartz and Gold:
The Summit in the Realm of Matter

There is a certain relation between gold and silica. That is even demonstrated purely quantitatively, because the higher the silica content of a mineral, the higher will be its gold content.

We have seen that silica is a substance most prominently involved in the formation of minerals and of the crust of the earth. Silica accounts for more than half of the mass of the crust of the earth. Gold, in contrast, we find in relatively small amounts in the earth's crust, although its presence can be demonstrated nearly everywhere. However, no less than silica, gold exhibits a strong affinity with earth. In compact form, it has one of the highest specific densities.

Silica we find on the whole in large, more or less opaque masses. In many places it shows its tendency to crystallize. It is possible to set up a whole graded sequence of transitions, beginning at one end with still half-formed, opaque, white, milky quartz crystals; at the other end we have crystal-clear, precisely-defined ones of "rock crystal." If we release the latter from earthy darkness and expose them to the sun, they give the impression of being created to receive the light—a fascinating play of light and color manifests itself.

In our discussion of gold, we saw how, in a manner of speaking, it also is predestined to be carried towards light, only in a totally different manner.

We could compare the way silica evolves in the midst of dull, stoney masses, into the most beautiful quartz crystals, and the way gold stands above the other metals, with the rising of ideals, lofty thoughts within our daily thinking. Rudolf Steiner points to gold in that connection.

Quartz is earth that raises itself so far that it can receive sunlight. Gold is earth that shines sun-like in the sunlight. Gold as matter can be extended nearly indefinitely without losing its shine. It behaves as if it wants to compete with light.

Quartz and gold represent the highest forms of the crystalline and metallic state that matter can achieve. Both demonstrate in a grandiose manner a reconciling of the contrast between gravity and light, between earth and sun. They teach us that everything terrestrial, if it is genuinely noble, will focus on and harmonize with the sun.[44]

Endnotes

[1] W. Pelikan, *The Seven Metals*, 2nd Ed (1993), Anthroposophic Press; ISBN: 0-910142-53-X.

[2] See Steiner, R., *First Natural-Scientific Lecture* (Light-course); GA 320, 9th Lecture.

[3] For a more detailed picture of the human being as described by Steiner, see *Theosophy*, GA 9; *Study of Man*, GA 293; B. Lievegoed, *Phases in the Human Being*.

[4] See Chapter III, "Salts, Acids, and Bases."

[5] Kolisko, L, *Sternenwirken in Erdenstoffen*, Stuttgart, 1936; More recent publications: Fyfe, A., "Moon and Plant, the imprint of Mercury/Venus/Uranus in the Plant Kingdom," capillary dynamolysis studies, 1967 to 1984—describing experiments which show a correlation between metal solutions and planetary positions. See also G. Piccardi, *The Chemical Basis for Medical Climatology*, C. C. Thomas, USA, 1962.

[6] Of course, this is the basis of homeopathic and anthroposophically-extended medicine.

[7] See ref. #1; W. Pelikan, *The Seven Metals*, 2nd Ed (1993), Anthroposophic Press.

[8] See "Tria Principia," Chapter I; also Chapter X, "Mercury and Tin as Metals."

[9] See ref. note 125 above.

[10] In the following table I have put together some data, taken from Greenwood and Earnshaw (note 46), relating to some physical properties of the metals to be discussed.

Physical property	Pb	Sn	Fe	Au	Cu	Hg	Ag
m.p.(°C)	327.00	232.00	1535.00	1064.00	1083.00	-38.90	961.00
b.p.(°C)	1751.00	2623.00	2750.00	2808.00	2570.00	357.00	2155.00
H_f/kJ/mol	4.81	7.07	13.80	12.80	13.00	2.30	11.10
H_v/kJ/mol	178.00	296.00	340.00	343.00	307.00	59.10	258.00
D_{20} g/cm^{-3}	11.30	7.30	7.90	19.30	8.95	13.50	10.50
M-ohm cm^{-1}	0.05	0.09	0.10	0.43	0.60	-	0.63

Hf = heat of fusion; Hv = heat of evaporation; D_{20} = density at 20°C.
All values for mercury are for the liquid state; the density of Sn is for the ß-modification which is stable at room temperature.

[11] We should be aware that Julius is talking about the property of reflection, or how silver and lead produce an image; however, if we look at the chemical reaction of both metals to the surrounding chemical environment, lead is initially more reactive, but most lead compounds are, chemically, very inactive in the long term. Silver is chemically inactive with respect to its chemical environment, with few exceptions, notably sulfur.

[12] See Chapter VII, "Carbon between Sun and Earth."

[13] See Chapter X "Characterization of the Metallic State."

[14] Julius is referring to domestic water, which is generally chlorinated. Rain and spring water will not have the same effect.

[15] See illumination apparatus, Chapter II, "Apparatus."

[16] Pelikan, W., *The Secrets of Metals*, Anthroposophic Press, 1984; see also Chapter VII, "An example of a Discussion of Sulfur."

[17] See Parkes, D.G., *Mellor's Inorganic Chemistry*, p 799; and Partington, J. R., *General Inorganic Chemistry*, p 525, MacMillan & Co. 1946.

[18] See Chapter VII "Alchemical and Mythological Viewpoints in Pottery and Porcelain Production."

[19] If we take the diameter to be 0.002 mm = $2 \cdot 10^{-6}$ meter, the length as 2.103 m, and the density as 19.4×10^3 kg/m^3, the weight of the wire should be only about 0.12 g.

[20] See Chapter VII, "Silica technology."

[21] See Chapter IX, "Fluorine, Chlorine, Bromine and Iodine."

[22] See Chapter X, "The Metallic State."

[23] Julius is indicating here how light influences form in plants and also how as one reaches the flower's fragrances, colors, essential oils, etc., these substances can be considered a refinement of the "crude" carbohydrates or starch; see, for example, Hauschka, *The Nature of Substance*, R. Steiner Press, ISBN 0 85440 424 45.

[24] See Chapter VII "The Occurrence of Magnesium Compounds in the Earth and in Living Organisms."

[25] Iron makes up 6.2% wt., is the 4th most abundant element in the crust, and the second most abundant metal; see Greenwood & A. Earnshaw, ref. 46.

[26] Kapp, E., *Grundlinien einer Philosophie der Technik* (Foundations of a Philosophy of Technology), Braunschweig, 1877.

[27] See, Baravalle, H. von, *Die Erscheinungen am Sternenhimmel* (The phenomena of the Starry Heavens)," 4th Ed., Stuttgart, 1962; also, Schultz, J., "Rhythms of the Stars."

[28] If all the iron were extracted from our body, we would have about 4 g of iron (about 0.005% of body weight), of which about 3 g can be extracted from hemoglobin (Mellor's p1275, ref. 45).

[29] See Chapter VIII.

[30] In the way iron has properties which we would only find among the 12 substances, aluminium has properties which make it possible to use it in engineering, as a heavy metal.

[31] See W. Pelikan, Ref. 121, p78.

[32] $Fe(OH)_3$ "contains" 45% O and no CO2; $Fe(HCO_3)_2$ "contains" nearly 50% CO_2 and 17% O; rewrite iron (II) bicarbonate as $Fe(OHCO_2)$.

[33] On rusting, see *Mellor's Inorganic Chemistry*, ref 45, p1250

[34] Goethe, J.W., *Theory of Color*, on the "Chemical Colors," 471-476 & 511.

[35] To give an example: iron can react with chlorine gas to form iron (II) or iron (III) chloride. The mass ratios are:

$$\text{iron : chlorine} = 56 : 2*35.4; \text{FeCl}_2 \rightarrow 35.4 : 56/2$$
$$\text{iron : chlorine} = 56 : 3*35.4; \text{FeCl}_3 \rightarrow 35.4 : 56/3$$

So with the same amount of chlorine, less iron reacts when producing iron (III) chloride than when iron (II) chloride is produced. This, of course, applies to all iron compounds. This is also an example of Dalton's law (see Chapter IV).

[36] Iron powder and sulfuric acid react as follows:
$$\text{Fe} + \text{H}_2\text{SO}_4 \rightarrow \text{H}_2 + \text{FeSO}_4$$

[37] See Partington (ref. 137) p 330.

[38] See this experiment, already described in Chapter II.

[39] See Chapter III, "Salt Formation from Acid and Alkaline."

[40] See Chapter II "Illumination from Beneath."

[41] See Chapter X, "Comparison with Substances in Element Circle."

[42] See earlier discussion of mercury, Chapter X, "Comparison with Substances in Element Circle."

[43] Mellor, ref. 45, p 289, mentions Stephen Hales as the inventor of the pneumatic trough and J. Priestly as the first to use mercury.

[44] See end of Chapter XII, topics "Silica as Oxide," "Silica Technology," "Quartz and the Sun."

XI

Completing the Whole Curriculum for Grade Twelve

Chemistry and the Human Organism

In the introduction to the first volume of this book it was indicated that it is not possible to develop a science which serves to any depth without starting out from the human being and consciously returning again to the human. We must start with the human being because—even if we are mostly unconscious of it—we ourselves pose the questions to which we strive to find answers. The form of the science we work out also will be decided primarily by us as human beings.

The converse is also true; it is one of the most important traits of humans that they pursue science. The welfare and development of humans depend to a large extent on science and the manner in which science develops. Our development is in large part thanks to the fact that we engage in science.

The sciences do not depend for their methods only on the human being but also on their subject matter. The field in which the researcher works also has a decisive influence on the method used.

If we consider the manner of research and especially the phenomena in the way indicated, we will find the human being is always and everywhere at the center, around which the world becomes organized. However, we can distinguish various layers in nature, and accordingly, different sciences have been developed which each relate, or at any rate should relate, to a specific aspect of the human being.

In physics, we not only use our sense organs, but the subject itself is largely a study of the sense qualities. The interaction between the human being and the world via the senses is one of the main research topics of physics, while to find a starting point for a well-founded mechanics, we should start with the human limb-movement system.

Conventional contemporary chemistry resembles physics too much, although we should take as our starting point a very different aspect of the human being. In the introduction to the first part of this book, it already has been indicated what we are dealing with here in the first place. Of course, we perceive many phenomena, which are occasioned through chemical interaction of various substances, with our senses. However, to find the true relationship of substance with our organism, we must descend deeper into its unconscious realms. In these realms, the human being is constantly confronted with the question: How can substances be brought into the sphere of life forces and be changed in such a way that they stand in the service of the whole organism? The organism must activate forces which are specific to certain substances. This specificity is only possible if substances have properties which make them "especially fit" to be taken up in the human organism.

Much material has been discussed in this book out of which comes to the fore the natural disposition of certain substances to be of service to the human organism. We have tried to develop a chemistry, which continuously manifests the relation to life, a chemistry that in its inner structure is an expression of the complex life forces in our bodies.[1]

Chemistry as an Intermediary between an Organic and Inorganic Science

If we now compare the path we have presented towards a "Goethean method" as described by Rudolf Steiner in his ***Theory of Knowledge***,[2] we then arrive at remarkable conclusions. If we work with chemistry as if it belonged solely to the realm of the inorganic, then the results are unsatisfactory. It is only possible to a limited extent to take the phenomena, and unravel and order them in such a way that we can trace them back to "archetypal phenomena." However, it is possible without further ado to discuss the chemical elements as if they were "types," which express themselves in ever new forms as we find in the realm of living nature. The question here is to develop a methodology that is intermediate between inorganic and organic science. Sometimes, we have to direct ourselves more towards "archetypal phenomena." At other times the processes occur as if they were being directed by something we can compare to a "type" in the plant world. An important rule was expressed in Chapters IV ("Guidelines for the Teaching of Chemical Formulae") and Chapter V ("The weight Ratio in Chemistry"). There, we stressed that for combining substances, their characteristics vanish— implying that they "merge into" one another and cease to exist as such, and new substances with totally different properties emerge. *This must be seen as one of the most characteristic archetypal phenomena of chemistry.*

In the chapters devoted to the properties and transformations of the various (elemental) substances, we encountered phenomena and thoughts arising which could appear to contradict the above point. For example, in the case of hydrogen compounds we discovered a tendency towards volatility and great mobility. In the case of sodium compounds we often encountered properties which encouraged solubility in water or self fusion.[3] However profound the transformation experienced by an element into its compounds, the "character"—or we could even say the "idea" of the element—even remains visible in the style of its expression and activity. The behavior of substances, seen from that point of view, reminds us of the "type" in the plant world. Every plant family is characterized by such a type. Thus, we have a lily-type, a rose-type, and a grass-type. Every variety within such a family is distinguishable from every other, and yet we clearly recognize the type that unites them, not only in the position of the leaves, the flower geometry, but also in properties which are harder to characterize, like their "manner of living," relation to the environment, etc. The family "type" (or, the "archetype" of the family if that is found clearly) runs through all sorts of forms, which appear as "varieties." We can only discover the relation between the varieties, if we can follow the type in their manifold metamorphoses. In this sense, we can speak of "metamorphosis" in the case of substances. However, here we encounter a difficulty in vocabulary. If we say: a chemical element can participate in a sequence of metamorphoses, then that is not correct. The element itself is a metamorphoses of the substance-type. Thus, we must distinguish between sulphur, which we can hold in our hands as a piece of matter, and the sulphur-principle, which is hidden therein. Perhaps we could speak of archetypal-sulphur (or primal-sulfur). The term "archetypal-sulphur" points to a dynamic which is hidden in the substance, and which expresses itself both in the properties of sulphur in elemental form just as much in those of its compounds. The world of substance is comparable to a tapestry, woven using the substance-types—archetypal-sulphur, archetypal-oxygen etc. The properties of compounds are determined through the manner in which the substances involved are woven and interact accordingly. In the case of the chemical elements, the properties are determined by one type. Therefore they often exhibit a one-sided, rigid form.

Thus, it is not the substance itself, with its physical and chemical properties, that is conserved under the most varied circumstances, but rather the "principle" in the substance, of which it is just an expression. We could nearly be tempted to speak of a "conservation of a substance style" (or "substance gesture"), in analogy to the law of the conservation of matter. However we articulate it, in actual fact we should take into account that chemistry is determined by two fundamental rules, which seem to contradict each other at first. For the more obvious properties of the substances, thus especially for those which manifest themselves immediately through the senses, the rule of disappearance and reappearance of properties in chemical analysis and combining, applies.

The other rule applies specifically to that aspect of the substance through which it relates to the entirety of the world. We are dealing here with the very special task, the special place which every element has, as carrier of this or that world principle. We are dealing with the inner dynamics of substance, which determine the style of its appearance and to which we can gain access rather more through our interrelating thinking activity, than through our senses.

Matter and Life

If we take consider the importance for life of various substances and in particular the chemical elements, it would seem as if substances could develop the characteristics they have "outside" of a living being as well as inside the being. Organisms utilise sodium compounds where liquids must circulate or where something must become a liquid,[4] and calcium compounds in those places where solid support or envelopment is necessary, and thus there where terrestrial forces have to be taken into account. Closer study, however, reveals something different. In as far as a substance is a carrier of a certain world principle, then it will indeed unfold this in the living organism. The living being has to avail itself of certain substances to be able to participate in these principles.[5]

Before it goes very far, the substance must be absorbed from a state which is totally alien to the organism. Plants absorb the substances directly from the mineral world. As soon as that has happened, those transformations take place which are only comprehensible if we assume that the normal inorganic properties of the substance have been suspended or even converted into their opposites. What sort of properties are we talking about? Chemical affinity, color, ability to melt, volatility, etc.,—properties which bring about interaction within the inorganic world. In contrast, in the living organism a totally different manner of interaction exists. Already when the substance enters the living being for the first time, it treats these substances as if common affinity does not exist. In a most grandiose way, transformations are brought about which would seem nearly impossible outside of the body. One of the most magnificent examples is photosynthesis, where carbon dioxide and water are converted into starch and oxygen.

Indeed, Rudolf Steiner speaks of the liberation of substances from inorganic forces, and a "tuning-in" to the forces which sustain life. He describes these forces precisely. One of their main properties is that in every respect they are opposite to inorganic forces. While the latter operate from the center, the others operate from the *whole* periphery, and, also, not just from a different point of the world periphery each time, but always from the whole circumference of the universe at the same time.[6] He calls these forces the "etheric formative forces."

Actually, with respect to life's processes we are dealing with *two* main processes. Initially, the substances are taken up, liberated from the inorganic play of forces, and taken into the domain of the formative forces. Slowly these then release the substances again, and they are given over to the mineral forces anew. However, in the meantime these substances have taken on those structures and properties which are typical of that particular living organism which has taken them up. These substances, thus, are not completely liberated from the living being, but remain in its service and bring about a suitable connection with the forces of the inorganic environment. These substances have likewise received its imprint. A clear example of such a substance which is on its journey back but has not given up its service, is the bark of trees. Something similar applies to wood.

Matter and Consciousness

Rudolf Steiner demonstrates in the work on healing that what occurs in the animal and the human being is more complicated than has been described so far. The same degree of tension and even contradiction exists between animal and vegetable, as between vegetable and mineral. The animal kingdom can only live at the expense of the plant kingdom. Furthermore, every animal must be able to conquer the vegetable, i.e., pure life, and place it in the service of a network of forces which is one level higher. In animals, all processes proceed in quite a different manner and according to different laws of in the plant kingdom. The substances which have been taken up not only have to yield or conform to the forces which give form and structure to life, but also to the specific animal forces.

In the human realm, the processes proceed in such a manner that the animal has also to be conquered, and the substances which have been taken in have to yield to a force-complex which stands yet one tier higher.

Life expresses itself in the plant kingdom predominantly by putting on and unfolding ever new forms. In the case of animals, growth is generally directed towards a sharply defined end-form. To make this possible, the tendency of the plant to create ever new forms must be restrained. The animal is formed in such a way that consciousness can light up to a certain degree, in conjunction with sense impressions.

Animals experience emotions, which results in certain holding-back. Animal nature expresses itself more in formed movement, than in producing certain forms. In animals, consciousness is entirely bound up with perceiving the environment, rather than containing perceiving processes within the organs and their own bodily activity.

What distinguishes human consciousness is that it can totally free itself from bodily processes. When our thought life operates in an undisturbed fashion, then its content and form is not determined by the body, but by the object it is contemplating. Thinking is *made possible by* the body, and although it is carried by it, at the same time it can be independent of the body. We should really consider this a miracle. To enable this to happen, the human being not only needs a well-balanced and upright human body, but also the typical animal processes must be overcome in its body, and elevated to a higher level. The digestion of the substances proceeds in such a manner that not only the plant-like is suppressed, whereby a consciousness bound to the body is occasioned, but also the animal is repressed to that extent that consciousness can free itself from the body in thinking activity.

Chemistry on Four Levels

If we ever want to understand how it is possible that the human spirit reveals itself through a material-body carrier, without being predestined in its activity by bodily processes, then we must consider what has been indicated above as a working hypothesis. Thus, we will not only understand the body as carrier of the human spirit, but also in this way reach an in-depth understanding of substance. Substance not only has inorganic properties, and it not only has the potential to be taken up and processed by plants, but it also has the potential to be used in the building-up of the animal, and finally of the human body. Only when this is clearly brought forward will chemistry take on a form through which it will be part of the all-encompassing world view, wherein the human being stands in the center of the world and the world is gathered around this being as its summary and center. Chemistry, thus, should not only be bio-centric, as Henderson[7] wants, but rather anthrocentric. As soon as we are able to distinguish the processes of substances in the human being as occurring at four levels, only then we will be able to develop a chemistry for the realms of nature, which will be complete within itself. Only then will substances show us their true nature.

There are indications of Rudolf Steiner wherein he considers the discussion of this fourfoldness of the planes of chemistry as a necessity for the curriculum of grade 12, albeit at a simple level.[8]

Victory over Materialism: the Task of Teaching Chemistry in Grade Twelve

Why is this such a necessity? Because only in this way can the student acquire a fundamentally consistent world-view, as we have already seen, and because it is difficult, if not impossible, to overcome materialism in the human being and in our culture without these insights.

What is the most important point that materialism actually deals with? Materialism holds the conviction that our thinking is completely dependent on our bodily processes, and that it is also a kind of extension of these processes. Ernst Haeckel (1834-1919), who later became one of the greatest propagandists for materialism, was a deeply religious person in his youth. During his studies, he attended the lectures of Rudolf Virchow (1821-1902), who tried with all means in his disposal, to prove that there is a causal relation between the brain structure and our thinking. He demonstrated, for example, that by removing specific parts of the brain, certain spiritual (mental) functions become unavailable. That was crucial for Haeckel. That should not be the case for students who have gone through our classes.

Overcoming materialism has for a long time not only been a question of world-view and belief. Medicine, for example, has increasingly taken on a form along the line which is based on the thought: we may, and even must, treat the human being as a machine. For example, we bring certain substances in the body, which we call medicines, and try to call forth certain effects, without taking the finer processes on these four levels into account. Thus, the human harmony is continuously being disturbed, and that implies a continuous danger of influencing the spirit and even binding it to the material processes. Nevertheless, only with a certain ordering of the structure of processes, can the spirit act free from matter.

It is no different with nutrition. We add chemical substances to our food, and we even try to produce nutrients in a purely chemical, artificial way. We interfere in many ways with the development of our cultivated plants by using fertilizers and poisonous herbicides or insecticides. Also, in this way, substances are introduced into our body, which strongly influence the delicate interplay of substances and forces.

The first thing which is necessary is understand that processes in the human being follow quite different pathways than outside of the body. The second thing we must be aware of follows directly from the first. We must develop a chemistry which takes into account the existence of the four realms of nature and which acknowledges four levels in the human. Both these points are the main consideration for the chemistry curriculum of grade twelve.

Only a first start has been undertaken in this book to elaborate what needs to be developed. Although there is some material which can be used to develop a "chemistry of four levels," the real problem, however, has not been addressed. I also believe that it is right to first work out the fundamental phenomena as clearly and transparently as possible, in their cosmic and human interrelationships, and only then to build further. But this building further, with respect to the human being, *must also* be taken on. Who will do it? Only those who prepare for the medical profession acquire enough usable material. Biologists and geologists do come into contact with a lot of phenomena which relate to the earth and to life in nature. However, they usually do not learn anything about the *human being*. Chemists have the most difficult fate; they must seek the path to the human being and to life more on their own than those previously mentioned, since during their studies they generally have received very little material in this area.

Even the medical doctor will need the greatest courage to travel this road all the way to the end, because the acquired ways of thinking will hinder them.

Thus, if we want to build a culture wherein materialism has been overcome right into medicine and nutrition, and wherein the human being fully comes to its own as a spiritual being, then those individuals who have been shown new pathways in their youth, in however rudimentary a form, will have to set to work.

Endnotes

[1] See the indications by Steiner in *The Origins of Natural Science*, GA 326; Dornach 1922/23, especially lectures 7 through 9.

[2] Steiner, R., *A Theory of Knowledge Based on Goethe's World-Conception*, GA 2, 1886.

[3] Julius wants to express that sodium (and potassium compounds) are the starting point for (chemical) weathering; the sodium salts are the first to dissolve and thus weaken the structure so that further disintegration occurs.

[4] See Chapter VII, "Lime (Chalk, Calcium)."

[5] Julius refers here to forces that act towards a center, or, as he says: operate from a center; they may be called "central forces"; the opposing forces, operate from the periphery and may be called "peripheral forces." One could also say: "gravity" forces, keeping matter within the sphere of the earth, the inorganic or mineral; the opposing forces, and forces of 'levity,' which are manifest in everything living, are responsible for plants pushing out of the soil. (See Steiner & Wegman, *Fundamentals of Therapy, an Extension of the Art of Healing through Spiritual Science*, GA 27).

6. See Steiner & Wegman, *Fundamentals of Therapy, an Extension of the Art of Healing through Spiritual Science*, GA 27.

7. Henderson, *The Fitness of the Environment*, see Ref.1; see also Chapter 1, "One-sidedness of Contemporary Chemistry and a Path to its Conquest."

8. See Steiner's grade 12 curriculum indications, April 30, 1924, p. 748 in *Conferences with Teachers*. Not a lot of research has been published in this area, although see also Klaus Frisch, "Aspects of Plant and Animal Chemistry" in *Erziehungskunst* July/Aug 1991, and Armin Scheffler, "Chemical Processes in the Kingdoms of Nature" in *die Drei* April 1982,